NON-FOOD USES OF AGRICULTURAL RAW MATERIALS

Economics, Biotechnology and Politics

Caroline A. Spelman

Research Fellow
Centre for European Agricultural Studies
Wye College
University of London
UK

CAB INTERNATIONAL

CAB INTERNATIONAL Tel: Wallingford (0491) 832111
Wallingford Telex: 847964 (COMAGG G)
Oxon OX10 8DE Telecom Gold/Dialcom: 84: CAU001
UK Fax: (0491) 833508

© CAB INTERNATIONAL 1994. All rights reserved. No part of this publication may be reproduced in any form or by any means, electronically, mechanically, by photocopying, recording or otherwise, without the prior permission of the copyright owners.

A catalogue entry for this book is available from the British Library.

ISBN 0 85198 769 9

Typeset by Solidus (Bristol) Limited
Printed and bound at the University Press, Cambridge

Contents

Introduction		v
1	**Antecedents of Non-food Uses**	1
	The Development of Society and Use of ARMs	1
	Diversification of Energy Sources	6
	The Interaction of Population, Land, Water and Energy in Agricultural Production	9
	Environmental Degradation	20
	Interdependence in Agricultural Production	24
	Conclusion	25
2	**Which Raw Materials will be Used?**	27
	Carbohydrates	27
	Oils and Fats	37
	Fibre Crops	44
	Wood	48
	Secondary Substances	52
	Energy from Waste Products	54
	Conclusion	55
3	**Technology**	58
	Biotechnology Defined	59
	Plant Biotechnology	61
	The Effect of Biotechnology on Process Technology	69
	The Impact of Biotechnology on Non-food Products	80
	Conclusion	82
4	**The Cost Equation: The Economics of Non-food Use**	84
	Competition between ARMs and NARMs	85
	The Development of Agricultural Prices	91
	The Underlying Economic Trend	92

	The Range of Non-food Uses	94
	The Economics of ARMs for Energy Use	98
	Macroeconomic Advantages of Using ARMs	105
	The Economics of Other Non-food Uses	108
	Alternative Forms of Support for Non-food Use	110
	Conclusion	113
5	**POLICIES AFFECTING NON-FOOD USES**	116
	Agricultural Policy	116
	Other Policies Affecting the Progress of Non-food Uses	129
	A Blueprint Policy for Non-food Uses	142
6	**CONCLUSION**	145
INDEX		150

Introduction

Less than a century ago, even in developed countries, people depended heavily upon the non-food use of agricultural raw materials (ARMs), notably as the feed for animals used for transportation. There was also heavy dependence on draught animals for tools such as ploughs and other agricultural equipment. Economic development and the progress of technology enabled alternative raw materials and processes to be used, and so in the developed world agricultural raw materials have largely been replaced by fossil resources for energy and transportation and their role restricted mainly to food uses. In less-developed countries, however, both food and non-food uses are still common. In Europe the limitation of ARM production for food uses has other causes. Military conflict resulted in a long period of food rationing after the Second World War and agricultural policies were designed to encourage food production. This is well illustrated by the objectives of the Common Agricultural Policy (CAP) of the European Community, which are set out in the Treaty of Rome of 1957. They reflect the desire to achieve security of supply for food. These agricultural policies have been almost too successful, resulting in over-production worldwide even if the surpluses are not evenly distributed. A new situation is emerging where the major agricultural producing countries in the western world face a choice as to what to do with land surplus to food production requirements. One possible alternative use for the land is to produce crops once again for non-food purposes.

Not only have there been shifts in the pattern of uses of agricultural raw materials over the centuries, but the range of crops and animal products has changed as different attributes were sought from the produce of the land. From a global viewpoint, the range of crops which can be used for non-food purposes is very large but in practice only a small amount of this potential is used on a commercial basis; this is due to technical, economic, biological and climatic constraints. New technology is, however, being applied to remove these constraints – for example, to adapt little-used crops to substitute for crops which can only be grown in certain climates.

The principal end-products sought for non-food use are carbohydrates, fats and fibres. This book explores not only the conventional crops used to produce these products, such as wheat, maize, sugarbeet, rape, sunflower, flax, hemp and cotton, but also less conventional crops such as lupins, elephant glass, and short-rotation woods, and agricultural by-products such as straw and agricultural waste.

Chapter 3 is devoted to the different techniques of the new biotechnologies, such as a recombinant DNA technology, which allow the wider use of ARMs. These techniques allow plants to be adapted to produce industrially tailor-made crops for non-food purposes, such as rapeseed varieties for the production of long-chain fatty acids. They have also revolutionized the fermentation industry, allowing not only improvements in existing fermentation processes to produce traditional products such as organic acids and alcohols, but also the creation of a new generation of products with attributes such as biodegradability. It explores the way in which traditional physical processes employing heat or chemicals can be replaced by biological processes.

The book takes issue with the prevailing view of the environmentalists that the expansion of a non-food agriculture could cause even greater damage to the environment. In fact, there are several ways in which non-food uses can make a positive contribution to environmental protection and conservation. Using genetic engineering it is possible to produce disease- and pest-resistant and nitrogen-fixing plant varieties which allow lower volumes of agrochemicals to be used (this applies to both food and non-food uses). The wider range of crops grown for non-food purposes will allow new opportunities for crop rotation and diversification, thus potentially enhancing biodiversity.

It is not just technological progress which has restricted the use of ARMs for non-food purposes in the past but also the economics of non-food production. In many cases, the ARMs have to compete with much cheaper substrates such as oil. But there is an underlying trend of the reduction in the price of ARMs relative to oil. Even before the 1990 Gulf crisis, ARMs had decreased in price to the extent that it was possible to buy four times as much cereal for the equivalent value of a barrel of oil as it was 20 years ago. Although these outlets may become viable in the long term, it will still be necessary to subsidize the use of ARMs for many non-food purposes in the short term.

The book examines the alternative methods used to support non-food outlets and those under consideration, such as the extension of the scheme for taking land out of food production but allowing crops to be grown on it for industrial purposes and the proposed carbon tax on fossil fuels.

There is a wide variation in the value of non-food products ranging from speciality chemicals, such as antibiotics where the price of the raw material is insignificant compared to the value of the final product through to bulk chemicals of low value where raw material costs are crucial. Two case studies are given to illustrate this point. One is the energy use of ARMs to produce a petrol additive ('bioethanol') for which the ARMs have to compete with products derived from oil. This product is made in large quantities in Brazil and the USA but in both countries it has to be subsidized. It is not yet produced in any significant quantity in the European Community. The second case study is of a biodegradable plastic made from ARMs but which still has to compete with oil-derived alternatives. There is a wide range of applications for this bioplastic, from high-value medical uses such as colostomy bags to medium-value uses as biodegradable packaging and wrappings.

Economic viability is not the sole criterion used by policy-makers to decide whether non-food uses should be encouraged. Other factors such as security of supply, balance of trade, employment and more recently the environment also have to be considered. Non-food uses offer benefits in all these cases. However, in the case of the environment, a delicate balance has to be struck between the methods used to produce the ARMs and the environmental benefits (or lack of disbenefit) that products made from them can provide. Potentially, new technology and biotechnology in particular can bring environmental benefits to both domains but a correct legislative framework is needed to encourage this to happen. The catalyst for change may be provided by the twin concerns of the need to liberate world trade, as expressed by the Uruguay Round of the negotiations on the General Agreement on Tariffs and Trade (GATT), and the desire to protect the environment as discussed at the Earth Summit in Rio de Janeiro in June 1992. The pressure to reduce agricultural price support may help to make ARMs more competitive for non-food use, providing farmers can remain in business in a more liberal trade environment.

Legislation varies from country to country where non-food uses have been developed. These are principally the USA, Japan and the European Community but also certain less-developed countries such as Brazil, and those of the Caribbean and parts of Africa where ARMs are grown for energy purposes. The agricultural, fiscal and industrial policies operating in these countries have shaped the growth of non-food outlets either to help or hinder their expansion. Examples are given of the US and Brazilian fiscal regimes intended to promote energy uses of ARMs. In contrast, an example is chosen from the EC which demonstrates how agricultural policy focused on food use may inhibit diversification, and recent efforts to remove these impediments are analysed.

The conclusion of the book draws together the issues presented in each chapter and assesses the changes necessary for non-food uses to expand on a large scale worldwide. There would seem to be a strong possibility that a two-tier pricing system for food and non-food outlets for agricultural produce may have to continue, at least for a transition period whilst non-food uses are being established. It argues the case for greater interdependence between countries for agricultural production. This would have two main benefits: firstly, it would allow those countries with a competitive advantage in producing certain types of crop to maximize the potential use of their resources and, secondly, it would provide a means of achieving a better distribution of agricultural produce worldwide such that shortages can be overcome and population growth, particularly in developing nations, can be sustained. Such a solution may prove impossible to achieve, however, on account of the lack of political will and the innate desire of nations for self-sufficiency in strategic commodities. There is also a danger of continued protectionism as more efficient non-food producers are prevented from entering the markets of countries trying to establish a non-food industry. The book speculates about the likelihood of the necessary changes being implemented to enable the unhindered development of non-food outlets and estimates a possible time frame.

1

ANTECEDENTS OF NON-FOOD USES

Over the centuries, humans have used agricultural raw materials (ARMs) for both food and non-food uses. The balance between the two has varied according to the availability of the raw materials and the level of development in the society that uses them. Historically, the primary use of agricultural raw materials has been for energy to grow food, protect health and improve the quality of life. The discovery of fossil fuels and the technology to recover and use them has reduced mankind's need to depend on agricultural raw materials for energy purposes. It is significant that most of the world's population living in developed countries is able to benefit from the availability of fossil reserves, whilst many in developing countries still have to depend on agricultural sources of energy. This contrast may not endure, since the Earth's fossil fuel reserves are finite and all populations will have to address their need to secure energy from a renewable source once again.

This chapter traces the development of agriculture which since the Iron Age has increased the cultivated area of the Earth to the point that there is now land surplus to food production in many parts of the world, such as Europe. It argues the case that agricultural progress will help provide solutions to the major problems of world hunger, population growth and dwindling fossil resources. It does not dismiss the arguments of the environmentalists against the extension of modern agricultural systems for non-food production but maintains that these should be seen in their correct perspective.

THE DEVELOPMENT OF SOCIETY AND USE OF ARMs

White (1943) proposed that the evolution of mankind occurred in three stages: (i) 'savagery' – people were hunter–gatherers living on wild food; (ii) 'barbarianism' – early agricultural and pastoral societies; and

(iii) 'civilization' – development of machines and intensive use of fossil energy to produce food and other necessities. White considered that man would have 'remained on the level of savagery indefinitely if he had not learned to augment the amount of energy under his control'. By the same token it could be argued that global economic development will be retarded unless a solution is found for the ultimate problem that fossil energy resources are finite. The following sections will examine the interaction between the history of society and its use of energy.

Hunter–Gatherers

Before the development of agriculture and formal crop production, wild plants and animals were the staple foods of humans and provided the main source of their raw materials. Clark and Haswell (1970) suggested that in favourable environmental conditions at least 150 ha of land would be needed to sustain each person, and in an only moderately favourable environment this figure would rise to 250 ha. Under more marginal conditions, such as those in a cold northern climate, as much as 14,000 ha per person might be needed. Obtaining food and collecting wood for fuel dominated the activities of these societies. The distances that had to be covered in search of these raw materials involved the expenditure of a considerable amount of time. In practice hunter–gatherers would not have scoured the entire area but concentrated on part of it at a time until supplies became exhausted and then moved on. It is estimated that a food provider in these societies would have spent on average 2.2 days per week collecting food, leaving 4.8 days for other activities such as gathering firewood, moving, constructing shelter, making clothing, caring for children and leisure. This breakdown is borne out by observing the practice of primitive bush societies today.

Hunter–gathering societies rarely exceeded 500 individuals (Bews, 1973; Lee and DeVore, 1976). Securing food and shelter consumed so much time that other activities were severely limited. This society could not support many dependants and was thus simple in structure. The development of agriculture made it easier and less time-consuming to secure food supplies, creating greater interdependence between people and a greater incentive for increased productivity. A food/energy surplus was planned for and this was used to support the non-farmers in society. An illustration of this comes from Egyptian society during the reign of the Pharaohs, from 2780 to 1625 BC. The availability of water from the Nile, the fertile alluvial soil and the warm climate allowed 5% of the population to be engaged in non-agricultural activities, notably the construction of the great pyramids by slaves (Cotterel, 1955). The Egyptian population remained fairly constant during this period at around three million. There is biblical evidence that the society was sufficiently developed to store food in case of famine, as Joseph did when he was placed in charge of Egypt and was able to provide food for his brothers' family from Canaan.

Early Agricultural Systems

The development of agriculture probably began by accident in hunter–gatherer societies as seeds dropped from materials brought back to the camps for different purposes. People would then have discovered they did not have to go so far to find the same raw materials. The relative ease of harvesting such crops at hand compared to random gathering would have encouraged deliberate planting. Early agricultural practice consisted of cutting back existing vegetation or burning it off to create a clearing, then planting and cropping for about two years. At the end of this time the plot would be abandoned as production declined as nutrients became exhausted and other problems such as pest outbreaks developed.

Animal husbandry may have begun when a hunter carried back the young of his prey to be reared by human hands at the camp. The first animals kept were probably chickens, ducks, pigs, rabbits, sheep, goats, cattle, camels, donkeys and llamas. They provided both food and a source of non-food materials such as hides, hair, horn, bone and feathers. Some of these animals would have bred in captivity, simplifying the work of the hunter yet again. Furthermore, the work of caring for the livestock could be done by the women and children, thus releasing the men for other activities necessary for the survival of the community. Maintaining herds of animals also gave greater security of supply. This would have been particularly important for people trying to survive in marginal habitats. The stabilization of the food supply also permitted a greater population to be supported on a given area of land than under the hunting–gathering system.

Some of the animals kept in captivity would have proved suitable for transportation, providing an additional source of power. The use of animal power to draw a plough was an immense breakthrough in agricultural production. One hour of oxpower could substitute for 3–5 hours of manpower. Horses were a significant improvement over oxen because they can move faster. About 3000 BC, the introduction of the wheel doubled the load of goods which could be transported by one person or one draught animal. The invention of the tractor provided yet another breakthrough: a 50-hp tractor can till one hectare of land in four hours, a task that would take a man 400 hours. All three sources of power – man, beast and tractor – require energy; a comparison of

Table 1.1. Comparison of energy inputs for tilling 1 ha of arable land by manpower, by oxen, and by 6-hp and 50-hp tractor.

Power	Hours	Energy inputs/kcal				
		Mechanical[1]	Fuel	Manpower	Oxpower	Total
Man	400	6000	–	194,000	–	200,000
Oxen (x 2)	65	6000	–	31,525	260,000	297,525
Tractor						
6-hp	25	191,631	237,562	12,125	–	441,318
50-hp	4	245,288	306,303	2400	–	553,991

Source: Pimentel and Pimentel (1979)
[1]This column represents the energy equivalent of the depreciation of the equipment over 400, 65, 25 and 4 hours respectively.

their energy requirements for the task is quite revealing (Table 1.1).

The advantage of animal power is that the ox can be persuaded to consume forage products which humans cannot use for food. In many regions forage can be considered a free energy source. Also, a waste product, like straw left from a cereal harvest such as rice, can be utilized by animals. Draught animals also provide a source of food as well as of power. Many of today's developed nations used draught animal power until comparatively recently. When the draught animals were providing power for agricultural production they would have consumed about 20% of the total agricultural production from the farm. Developed countries have now largely replaced draught animals with machines. In the USA, manpower was the primary source of power in the early settlement period from 1620; by 1776 an estimated 70% of power was supplied by animals and today this has fallen to less than 1%, while fossil fuel engines account for 95% of power.

The Development of Agriculture

The development of agriculture is described by Buringh (in Pimentel and Hall, 1989) as having gone through seven stages characterized below:

1. Land rotation or shifting cultivation or bush fallow (a few years of crop cultivation followed by 10, 20 or more years of regrowth of forest).
2. Low traditional (mainly one crop year and one fallow year).
3. Moderate traditional (mainly two crop years and one fallow year).
4. Improved traditional (continuous cropping of cereals in rotation with legumes, root crops or grass).
5. Moderate technological (continuous cropping with applications of

some chemical fertilizers and simple mechanization).
6. Highly technological (using more fertilizers than the preceding, and with increased mechanization).
7. Specialized technological (similar to the preceding, but with heavy applications of pesticides).

Each of these successive stages employs more off-farm inputs and shows an increasing dependence upon fossil fuels, either directly or indirectly. As soon as technological modes of production are used external inputs become essential to the system. This raises the question of what will happen when fossil resources prove more difficult to obtain. For the same level of production to be maintained alternative production methods employing non-fossil inputs will have to be sought. The purchased inputs are not the only inputs in the agricultural system; as the level of development progresses through the seven stages some inputs per unit of output fall whilst others rise, but the total level of inputs (fossil and non-fossil) usually falls as productivity rises.

The use of commercial energy in the developing countries is only about one quarter per unit of agricultural production of that in the developed world, and this differential is reflected in the lower yields obtained by developing countries. If farm yields and earnings are to rise in these countries, there must be an increase in their use of commercial energy. The Food and Agriculture Organization has made an estimate of the extent to which energy consumption could rise in 90 developing countries by the end of this century (Table 1.2).

Table 1.2. Projected percentage changes in commercial energy use in agriculture in developing countries.

	Fertilizer	Machinery	Irrigation	Pesticides	Total[1]
90 countries:					
1980	54	31	12	3	36
2000	57	38	4	2	178
Low-income countries:					
1980	59	16	22	3	14
2000	69	24	6	1	91
Mid-income countries:					
1980	51	41	6	3	22
2000	44	52	2	2	86

Source: FAO (1981)
[1] Million tonnes oil equivalent.

The projections in Table 1.2 are based on the assumption that the agricultural systems of developing countries will become steadily more energy-intensive, with each 1% growth in output requiring 2.4% growth in the use of energy. This need for developing countries to become more energy-dependent as finite fossil energy reserves are being depleted may seem paradoxical, but it is essential if these countries are to develop economically. At the 1992 Earth Summit in Rio there appeared to be little political will amongst the developed nations to conserve fossil energy, or to consider priority being given to the needs of less-developed nations of these resources to improve their own level of development.

DIVERSIFICATION OF ENERGY SOURCES

Historically, in primitive agricultural societies by far the greatest amount of time had to be spent on food production and it was not until draught animal power was introduced that more time was liberated for other activities. There then followed great progress in the diversification of energy resources, including wind and water. With these changes came greater diversity in people's activities and an increasing degree of specialization into farming, sailing, trading and industry. The importance of food production in primitive society can be assessed from the fact that 95% of total energy was expended on it. In some developing countries today the proportion of energy expended on the food system is still 60–80%. By contrast the proportion of energy expended on food production in a developed society such as the United Kingdom or the USA represents about 16% of the total energy used (Leach, 1976). The main explanation for this massive reduction in the amount of time, energy and manpower required for food production in a developed society is the use of increasingly concentrated forms of energy.

The use of energy is governed by the laws of thermodynamics. The first of these states that 'energy may be transformed from one type to another but can never be created or destroyed'. So, for example, light energy from the Sun can be transformed by plants into food (chemical) energy, which in turn can be transformed by human beings or animals into capacity to work. The second law of thermodynamics is also relevant; it states that 'no transformation of energy will occur unless energy is degraded from a concentrated form to a more dispersed form', and further that 'no transformation is 100% efficient'.

The function of all biological systems follows this second law of thermodynamics when solar energy is converted to chemical energy. In the process of this conversion, some of the energy of sunlight is lost as heat to the surrounding environment. The survival of humans and their ecosystem depends on the efficiency of green plants as energy converters. In fact the amount of energy fixed by the world's plants amounts to less than 0.1% of the total sunlight reaching the Earth.

The breakthrough in the use of animal power and fossil fuels becomes apparent if their relative concentrations are compared. Horsepower is a commonly used unit of power. One horsepower-hour (hp-h) is the capacity to do 33,000 foot pounds of work per minute for one hour, and is based on the ability of the average horse to lift 33,000 pounds one foot per minute for one hour. The maximum daily work capacity of a horse is about 10 hp-h or a ten-hour work day. By comparison, one man working a ten-hour day produces only the equivalent of 1 hp-h. Thus one horse in one hour can accomplish the same amount of work as ten men in one hour. The tremendous concentration of energy in fossil fuels can be appreciated by comparing them to manpower. One gallon of petrol contains 31,000 kilocalories of potential energy. When used in a mechanical engine which is about 20% efficient at converting heat energy into mechanical energy, it produces the equivalent of 9.7 hp-h of work. Thus from one gallon of petrol it is possible to obtain the same amount of work as from a horse working at capacity for a ten-hour day.

Shifting to the Use of Fossil Fuels

The catalyst for the shift to the use of fossil fuels was the increasing shortage of wood available to use as the main fuel. By the 16th century in England wood was becoming scarce and soft coal was sought as a substitute. As mines were dug deeper flooding became a problem, which stimulated the development of the first steam-powered pumps at the end of this century. It was not until 100 years later that James Watt designed a reasonably efficient steam engine and pump. The USA continued to use wood for longer on account of its greater availability in that country: as late as 1850, more than 91% of energy used in the USA was from wood. The Watt steam engine and the internal combustion engine radically transformed the use of energy. Today, consumption of fossil fuel is at its highest level ever, and increasing. Fossil fuels are the source of 94% of all energy consumed in the United States.

This period of dependence upon fossil fuels is likely to be viewed as a 'blip' in the pattern of energy usage in the history of mankind. In the centuries of documented existence of humans on Earth the reliance upon fossil fuels is likely to be limited to 500–600 years on account of the finite nature of their reserves (Table 1.3). Oil and gas supplies will be the first to be exhausted; best estimates are that more than half of these fuel reserves will be used by the end of this century (Pimentel and Hall, 1989). Coal reserves are more extensive and expected to last for the next 100 years.

Table 1.3. Annual production and reserve/production ratios (R/P) for oil, coal and gas 1970 and 1989.

	1970		1989	
Fuel	Production	R/P	Production	R/P
Oil	16.7 bn barrels	31 years	21.4 bn barrels	41 years
Coal	2.2 bn tonnes	2300 years	5.2 bn tonnes	326 years (hard) 434 years (soft)
Gas	30 trn cu.ft	38 years	68 trn cu.ft	60 years

Source: Meadows and Meadows (1992)

Table 1.4. Global distribution of energy use/%.

	Total	Developed	Developing
Biomass	14	1	43
Coal	27	28	26
Oil	39	45	24
Gas	17	23	4
Hydro/nuclear	2–5	3	1.5

Source: Hall et al. (1982)

Coal, however, produces sulfur dioxide when it is burnt and this gas is considered to be one of the main causes of acid rain. All the fossil fuels suffer from the disadvantage that they contribute to the volume of carbon dioxide in the Earth's atmosphere when burned; this is widely expected to give rise to the so-called 'greenhouse effect' or global warming, which may upset the delicate ecological balance upon which life on this planet depends. Therefore there may be reasons why the use of fossil fuels should be curtailed before reaching the point of exhausting the supplies.

There is a startling contrast between the types of energy used in developed and developing countries (Table 1.4). As the society becomes more sophisticated it depends less and less on biomass for energy. The substitution of biomass by fossil fuels has enabled greater efficiency to be attained, as well as greater economies to be achieved. For this reason the standard of living achieved from most man-powered systems is relatively low compared with that possible when large inputs of fossil fuel are used in crop production. Thus fossil energy has helped mankind to manipulate ecosystems more effectively and efficiently for food production than ever before, and has contributed to a higher standard of living in those countries that use them. The wheel may turn full circle again, however, as reserves of fossil fuels are used up and biomass as a renewable energy source is given fresh consideration.

The Future Use of Biomass

As potential energy sources, terrestrial and aquatic plants offer environmental advantages over fossil energy. One of the most obvious and topical is the fact that plants absorb carbon dioxide in the process of photosynthesis as part of the natural carbon cycle. They do not aggravate the so-called 'greenhouse effect' and may even help to reduce it. They renew the oxygen supply and help remove chemicals from the atmosphere which might otherwise cause the development of a harmful photochemical smog damaging the ozone layer. The ozone layer prevents a large percentage of the Sun's ultraviolet light from reaching the Earth. No terrestrial life as we know it could exist without this protective shield. This is the basis of the recent concern about the reduction in the ozone layer. Another vital function performed by plants is the breakdown of wastes produced by man, agriculture and the natural system itself. Bacteria, fungi, protozoa and invertebrates all help degrade wastes. These reducing organisms all help recycle essential minerals for re-use by other members of the ecosystem. Primitive humans did not realize or understand these environmental benefits, but today we both understand and make use of them.

The combination of environmental advantages of biomass, particularly renewable plant material, coupled with the shrinkage of finite fossil fuel reserves has helped to prompt interest in renewable energy sources derived from agriculture. One of the principal objections to this has been that in some developed countries the diversification of agriculture away from food production is taking place while some parts of the world are still short of food. A second major objection has been the possible negative environmental consequences of large-scale biomass production.

In the next section, the capacity of the world's resources to meet both demand for food and non-food products will be discussed.

THE INTERACTION OF POPULATION, LAND, WATER AND ENERGY IN AGRICULTURAL PRODUCTION

One of the earliest documented views of humans and their role in their ecosystem is found in the first book of the Bible, in Genesis 1:28, which says 'Be fruitful and multiply and replenish the Earth and subdue it.'

Mankind has undoubtedly been successful in multiplying in numbers and in the subduing part of this injunction, but not until recently in the replenishing. The development of tools and machines, coupled with the discovery of such new sources of power as fossil energy and nuclear energy has enabled humans to exert tremendous control over their environment. The challenge now facing us is how to make optimum use of the remaining resources and to sustain an ever-increasing population. Environmentalists have delivered dire warnings of the dangers of making the wrong decisions now or of not heeding signals about damage to the environment. These are well illustrated by the views of Denis Gabor of Imperial College, London, which predate the birth of the environmental movement. He said, 'exponential curves grow only to infinity in mathematics. In the physical world they either turn round or they break down catastrophically. It is our duty as thinking men to do our best towards a gentle saturation instead of sustaining exponential growth, though this faces us with very unfamiliar and distasteful problems' (in Forbes, 1968). Here we shall assess the potency of the environmentalists' views and discuss what could be done to avoid a potential catastrophe.

Population Size

A great debate rages about the size of population the Earth can sustain. The present rate of population growth of almost 2% per annum is comparatively new in human history. If maintained, it will lead to a doubling of the world population every 30–40 years. For comparison, it took the world population three centuries to double between 1500 and 1800. During that time the actual size of the population rose and fell in waves according to the impact of epidemics, food shortages and wars. There is a current popular view that at present the world, with 5 billion inhabitants, is over-populated. Notwithstanding the importance of family planning to control population size, world leaders at the Earth Summit held in Rio in 1992 failed to endorse efforts to improve on managed population growth. Those in favour of wider use of family planning justify their views on the grounds that lack of food is a major problem in certain parts of the world. So, the question arises as to whether the Earth has reached a threshold of food production? Calvin (1986) made an interesting calculation of the maximum amount of food that could be produced from the Earth. He began with the assumption that the total amount of solar energy reaching the Earth is $5.5 \times 10.16 \, \text{kcal day}^{-1}$. He multiplied that by the proportion of the Earth's surface which is arable crop land (20%) and multiplied the product in turn by the photosynthetic efficiency of plants (0.1%). He then divided the result (the energy available each day to the Earth for food production) by the average daily energy requirement of a human being, 2500 kcal. This calculation showed that in theory the world could support a population of 22 billion.

Another way of getting to a theoretical maximum population size related to the availability of food is to examine the potential to improve yields of existing crops. If the present yields of grain achieved by the developed world could be replicated wherever grain is produced, there would be enough grain to support twice the present world population of 5 billion. Through improvements in plant breeding and biotechnology further yield improvements are likely to be obtained. In the laboratory it is possible to quadruple even existing grain yields from developed countries such as the United States. This corroborates Calvin's argument that a population in excess of 20 billion, or four times the present size, could be supported.

It must be acknowledged that even at the present population size there are many people who do not have enough food, but it is equally wrong to extrapolate from the cases of those suffering from food shortages that enough cannot be produced. It is a fact that world grain production has doubled in the last 20 years and *per capita* consumption has also increased, even if these improvements have not been evenly distributed throughout the world. These improvements could be even greater, but ironically production has progressed least in those countries which could produce more on account of the over-supply of world markets. Farmers in the USA and Europe are actually being paid not to produce. It can sound trite to say that world hunger is a function of the distribution of resources and not an absolute shortage, but it is true. It has been estimated (Hudson in Pimentel and Hall, 1989) that enough grain could be produced on land retired from food production in the USA to feed 250 million people. These facts negate the objection to deploying the Earth's agricultural resources for non-food uses.

Another objection relates to the environmental damage which could be caused by the extension of agricultural production. It is accepted that water, labour, land and energy can only be substituted for one another within certain limits in the management of agroecosystems. Land quality affects the yield of crops and can be improved by the use of more fertilizers, water and other energy inputs. There follows an assessment of the availability of these factors.

The Availability of Land

It is difficult to assess the amount of land available for agriculture. Data presented to organizations like the Food and Agriculture Organization (FAO) are at best estimates. It is believed that 10% of the total world land area is covered by ice, 15% is too cold and 17% too dry to grow crops, while 4% is too wet, 9% too rocky or stony and 18% too steep and 5% too poor for other reasons; that leaves only 22% suitable for crop production. In theory, a total area of 13.4 billion ha is considered available for agricultural use, of which 20% is suitable for arable crop production. However, only about half of this is actually used. Therefore there is still some land which could be brought into agricultural crop production although account would need to be taken of practical and economic constraints.

There is also considerable potential to restore the lands rendered unproductive by past mismanagement. There are now an estimated 2 billion hectares of degraded land which could be brought under cultivation. These include land degraded by non-agricultural uses, as well as land abandoned on account of its declining agricultural productivity. For example, there are an estimated 500 million ha land reserves in the humid tropical regions of Latin America, Africa, and eastern Asia which could be used for arable crops. Although it would be politically unacceptable to destroy the virgin rainforests of these regions in order to establish new sites for the cultivation of non-food crops, biomass production offers a means of recovering, restoring, sustaining and enhancing the productivity of these neglected lands (Lowry, 1986).

This assessment is complicated by the fact that each year land is lost from production because of erosion, salinization, degradation and non-agricultural development. Buringh (in Pimentel and Hall, 1989) estimates that 4 million hectares are lost annually. The erosion of topsoil is an example of such a problem. Topsoil depths average 18–20 cm. For each 2.5 cm lost the reduction in crop productivity has been estimated to be 251 kg maize, 161 kg wheat, 168 kg oats or 175 kg soyabeans per hectare (Pimentel *et al.*, 1976). When primitive agriculture was being established (about 10,000 years ago) fire was used to help clear trees and shrubs from the crop land; this simple procedure helped to eliminate weeds in the absence of modern herbicides. The incorporation of ash in the soil added nutrients and made the land more productive. This slash-and-burn or 'swidden' technique continues today in some less-developed countries, and in particular where there is difficult terrain for cultivation. However, it has aggravated the problem of soil erosion as well as reducing the world's virgin forests. If the correct technical advice and assistance could be given to people living in these areas, unnecessary soil degradation could be prevented. Soil erosion is as, if not more, serious in developed countries, particularly where land is intensively farmed.

There is a massive difference in the land use policies of the developed and developing world. The developing world still struggles to provide enough food for its indigenous population, let alone to produce enough cash crops for export to obtain revenue for necessary imports such as fossil energy. By contrast, in large parts of the developed world land has been taken out of food production to maintain farmers' prices, to manage domestic markets and to alleviate the depressing effect that exporting the surpluses had on world markets. Fresh impetus has been given to this initiative by the agreement reached at the Uruguay Round of the negotiations on the General Agreement on Tariffs and Trade (GATT) which requires further reductions in exports by the developed countries, thus implying an increase in the land available for alternative uses. The developed world enjoys the luxury of choice in the use of land retired from food production. It can be allowed to lie as fallow land, 'set aside' or put to an alternative agricultural or non-agricultural use. There are usually quite strict legal conditions on change of use from agricultural to non-agricultural purposes. Thus an easier option legally is to convert the land to non-food production such as biomass for energy.

The Availability of Water

Water is a scarce resource. Usable water represents less than 0.7% of global water resources. The greatest quantity of water exists in oceans (97.13%) and the second greatest quantity as ice and snow in the polar regions. Ground water represents 0.612%, lakes 0.009%, and rivers and streams 0.0001% of the total water resources of the world. The water resources are not evenly distributed and agricultural production is only sustained in some parts of the world by irrigation. The world average amount of land under irrigation is 14.5%, but some continents have a far higher percentage than others. Asia, for example, has 29.3% of land under irrigation and accounts for the largest proportion of irrigated land in the world: 62.5% of the total. In Europe and in North and Central America only about 10% of the land is irrigated. The availability of water depends on the hydrological cycle of precipitation, evaporation, transpiration, infiltration, percolation, run-off and storage. Complicated systems have been developed to maximize the water resources available, including the pumping of ground water, artificial damming and desalination and recycling, but these are not used to their full potential and there is still an unacceptably high level of water wastage.

The availability of adequate water supplies often influences the amount of land needed for crop production. With ample moisture and fertilizer use, crop plants can be grown densely and high yields produced. The choice of crop grown by the farmer can be influenced by the efficiency of the plant to make use of irrigated water. This in turn is influenced by factors such as foliage cover of the soil and the transpiration characteristics of the crop. The economic efficiency of irrigated crops varies (Table 1.5; this table is drawn from a statistical

Table 1.5. The economic efficiency of irrigated crops.

Crop	Water applied /m³ ha⁻¹	Crop yield /t ha⁻¹	Crop value /$ t⁻¹	Water use /m³ ha⁻¹	Water use /m³ $⁻¹
Alfalfa hay	18,000	12.2	93.1	1474	15.8
Maize	9000	7.7	126.0	1164	9.2
Rice	18,300	1.1	1289	10,108	19.8
Potatoes	9900	37.1	179.4	267	1.5
Sugarbeet	13,500	56.5	38.7	239	6.2
Wheat	6000	5.5	127.0	1097	8.7

Source: University of California (1977)

report of the University of California on water but shows only crops of interest to the subject of this book).

The data shown in Table 1.5 are a function of crop yield, market price, water requirement and management practices. They indicate that some crops are more efficient users of water than others; rice and alfalfa being notably less efficient with water than potatoes, sugarbeet, wheat and maize.

Agriculture can also be affected by water quality, and farm practices can themselves cause water degradation. Water quality can be adversely affected by salts, nitrates, residues of agricultural chemicals and soil erosion, as well as slurry from intensive livestock production and waste from food processing. More care could be taken to improve the quality of water by modifying agricultural practices themselves. The first steps have been taken in some parts of Europe with the designation of zones for water protection (*Wasserschutzgebieten* in Germany, nitrate-sensitive areas in the UK).

Salinity can be a particular problem in some areas and precludes the use of otherwise suitable land for arable crops. Some crops have a greater degree of tolerance to salt than others; examples of these are sugarbeet, date palm and cotton. There are also some specific types of wheat which will grow in salty soil. These crops will not tolerate excessive levels of salt without some loss of yield, but there is scope for improving salt tolerance by genetic engineering.

There has been much speculation about global climatic change and its origins. There seems to be evidence that droughts are becoming more frequent even in temperate climates, and it is postulated that this is one outcome of the greenhouse effect. Notwithstanding any climatic changes that there may already have been, world grain yields have risen steadily, and indeed quite steeply in some parts of the world. Other parts that have traditionally seen droughts are faced with an extension of drought-prone zones, making agricultural production even more precarious. Two thirds of Africa receives more than half its rainfall in only three months. There is also great variation in rainfall from year to year, which results in crop failures. The drought-prone area of sub-Saharan Africa is expanding and includes the countries of Chad, Niger, Sudan, Ethiopia, Mali and Mauritania; further south, Angola and Malawi are also critically affected. Hunger is a primary problem in these areas, but crop failure also affects the non-food needs of the population for raw materials for shelter and clothing. Skilful planting of specific crops and trees in these areas can help ameliorate water deprivation and its consequences. There are various current success stories in agroforestry, including the growing of food and non-food crops between regularly pruned hedgerows of fast-growing, nitrogen-fixing trees; the trees themselves serve multiple purposes, acting as windbreaks to reduce evaporation and for the production of fertilizers from leaves and twigs used as a mulch (Falkenmark, in Pimentel and Hall, 1989). The overall picture is thus one of increasing crop yields, even if the yields in a particular country may fluctuate considerably from year to year.

The Availability of Energy

The world currently consumes approximately ten times as much energy as it did 100 years ago, equal to 150 million barrels of oil equivalent per annum. Historically, wood was replaced as the primary energy source by an enormous rise in coal consumption before the First World War after which coal gave way to a steady growth in oil usage. After the Second World War, a market was established for natural gas whilst wood and coal use continued to decline. The use of nuclear energy did not really take off until 1960, and even since then its growth has been tentative on account of the environmental risks it poses. Today 55% of energy demand is met by oil, 25% by coal, 12% by nuclear energy and 8% by hydro and other energy forms.

The pattern of demand for energy shows contrasting trends between developed and developing nations. Over the last 20 years the developing countries' share of world energy consumption has risen from 14 to 25%. This rise differs markedly from the relatively slow growth of demand for energy in developed countries. The relationship between economic activity and energy demand is expressed as a country's energy intensity, the ratio of energy use to gross domestic product (GDP). There is a general decline in energy consumption amongst developed nations on account of the development of less energy-intensive postindustrial service

Table 1.6. Average annual growth rates of primary energy demand per unit of GDP, 1975–1985/%.

Low-income countries:		Lower middle-income countries:	
Pakistan	+4.2	Nigeria	+9.4
Senegal	+3.6	Ivory Coast	+2.8
India	+1.4		
Upper middle-income countries:		**Industrial high-income countries:**	
Venezuela	+4.6	Australia	–0.5
Mexico	+2.2	Germany	–1.7
Brazil	+2.1	Netherlands	–1.7
Argentina	+1.8	USA	–2.2
		Japan	–3.1

Source: World Resources Institute (1990)

economies and the introduction of energy-efficient conservation measures. By contrast the developing countries of Asia, Africa and Latin America are generally increasing their energy consumption rates (Table 1.6).

The worldwide pattern of distribution of fossil energy reserves is uneven, with a concentration of reserves in certain areas and many countries having to depend completely on imports of fossil energy. The Organization of Oil Producing Countries (OPEC), whose major members are Saudi Arabia, Iraq, the United Arab Emirates, Venezuela, Libya and Nigeria, dominates the reserves of oil. The countries of the former Soviet Union account for the largest reserves of natural gas. Canada and Venezuela have vast reserves of tar and bitumen and China the largest coal resources. Not all of these are recoverable. Table 1.7 assesses the world reserves of recoverable fossil energy and their approximate cost.

Table 1.7 shows the oil industry's estimates of the extent of reserves and their approximate cost. The unit technical cost certainly does not represent the market price. The cheapest energy produced is the light oil of the major OPEC producers, which would have to be sold at not less than $10 per barrel to cover current and private costs. Other producers struggle to produce oil at half as much again, and tar sands and oil shale are just not economic to use at current oil prices (hence the indefinite time scale of their use). The time it would take to exhaust these reserves, and the question whether they ever will be exhausted with current costs, are both disputed.

Since the extent of the remaining fossil energy reserves is disputed, it is perhaps revealing to consider energy availability in relation to consumption patterns in the developed world's population. It is estimated that 17% of total energy use is expended on the production

Table 1.7. World reserves of non-renewable energy.

	Approximate recoverable reserves[1]	Reserves	
		Production years	Unit technical cost[2]
Major OPEC members	800	100	3–10
Other oil producers	500	45	4–30
Gas	1500	125	1–30
Coal	5000	centuries	5–40
Heavy oil	500	?	15–30
Tar sands	500	?	20–40
Shale oils	1000	?	25–40

Source: Jennings (1989)
[1] Billion barrels oil equivalent.
[2] $ per barrel of oil equivalent.

of food in developed countries: that is, about 1500 litres of petrol *per capita* each year. If this is extrapolated to the entire world population of 5 billion, it would mean 7500 billion litres. The known world petroleum reserves amount to 113,700 billion litres. Assuming only 76% of that can be converted to fuel, and that all of this is used for food production, the reserve would last only 12 years even with no increase in population.

Since all fossil energy resources are based on carbon, their burning aggravates the greenhouse effect through the production of carbon dioxide. There is a distinction, however, between different types of energy as to the extent to which they are polluting. Oil produces 15% less carbon dioxide (CO_2) than coal, and natural gas 30% less CO_2 than oil. Moreover, carbon dioxide is not the only component from burning oil which causes harm to the environment; sulfur dioxide and nitrogen oxides are also harmful, though they are easier to isolate. Because of their complex structure, fossil fuels are likely to cause an endless stream of concerns about the environment and health. Attention has now moved on to the possible carcinogenic effects of the aromatic components such as benzene. Stricter environmental standards may accelerate the decline in the use of fossil energy as other, less harmful forms of energy are preferred. One way in which this might be achieved is through the introduction of a carbon tax on fossil fuels.

This proposed environmental levy on petroleum and other hydrocarbons is intended to curb their use and reduce carbon dioxide emissions to help prevent global warming. The European Community has initiated the proposal for a tax which would commence at $1–3 per barrel of oil in 1993 and rise to $10 per barrel by 2000. Already, some of the Scandinavian countries have implemented such a tax. In Norway, it represents 12.6 US cents per litre of petrol. Italy has proposed that a $1 per barrel levy should be adopted by all 24 nations of the Organization for Economic Co-operation and Development (OECD), but this was not accepted at the Rio Summit meeting in June 1992. The carbon tax is controversial, for whilst it would raise considerable sums of money for the exchequers of the developed nations, it would hit the oil-producing nations hard without giving them any means to diversify away from oil production. It is also controversial within the developed oil-using countries themselves, because it could drive up the cost of energy of domestic industries who could not then compete with the industries of nations choosing not to implement the tax (*Wall Street Journal*, European edition, 9 June, 1992).

The estimate of the different types of energy consumed at the start of this section included 8% for hydroelectric and other types of renewable energy. This relates to commercial energy demand. Renewable energy is estimated to satisfy 20% of world demand for energy if non-commercial sources are included. This reflects the high dependence upon biomass fuels in developing countries. Other renewable sources of energy such as hydroelectricity, wind energy and solar energy tend to be developed at times when world prices of non-renewable energy rise and concern about security of supply increases. These less conventional sources of energy have also found niche markets for systems and products which are not solely related to oil prices. Hydro-power is currently used to generate 25% of the world's electricity. The potential for the development of hydro-power is limited by the number of suitable locations and the considerable local environmental impact. The use of solar energy, particularly for domestic water heating, has grown remarkably in the last 15 years. There is also a growth of interest in simple solar cookers for developing countries. Prospects for the development of high-temperature solar units for steam and power generation are much longer-term, and it is more likely to supplement conventional fuels in a combined system. Modern wind energy systems can be competitive with conventional generating facilities in areas with an adequate wind supply. Future potential will be influenced by views about the impact upon the environment of these 'wind farms' (noise, visual effect) but principally by the comparative cost of other forms of energy. Considerable opportunities already exist for using renewable energies in combination with conventional sources to provide energy-efficient systems in a conservation-conscious world energy market. The fundamental opportunity for renewables remains the same, however: they will inevitably play an increasingly important role, with growing energy demand and ultimately limited hydrocarbon reserves. A most important issue is the rate at which these finite reserves will be exhausted, but this is difficult to estimate on account of the continual discovery of fresh reserves as a result of oil exploration.

The Availability of Forest Resources

Since grasslands and forests constitute 85% of global biomass production, they represent an important agricultural resource. Wood has the great advantage of not necessarily competing with food crops for land and water resources. However, this does not mean there is room for complacency. Worldwide deforestation is removing about 11.6 million ha of forest annually. Most of this is due to the demand for timber, but soil erosion through agricultural practices and loss of soil fertility are other important causes. Demand for 'new' agricultural land in less-developed countries will put even more pressure on the world's forest resources, which are thus themselves a constraint on development. It is important that these forest resources should be used more judiciously, especially in developing countries. Forests are essential for future economic development as they provide lumber for housing and other products, pulp for paper, and biomass for fuel. There is a fundamental difference between developed and developing countries in that the latter depend on wood for fuel. In the Far East, fuel wood accounts for 42% of total energy use and in Africa this figure rises to 58% (FAO, 1981). This figure is an average and would be much higher for the poorer of these countries. In addition organic materials which should be returned to the soil, such as crop residues and dung, are increasingly being burnt as the 'poor man's solution' to the energy crisis.

There are means of easing the constraints on forest resources with modern agricultural techniques. For example, a well-managed village wood plot of fast-growing tree species can yield six times the wood quantity of unmanaged forest (FAO, 1981). The situation would also be improved if the wood were burnt in a more efficient manner, i.e. in efficient wood-burning stoves rather than inefficient open fires. A family of six would then only need about 0.25 ha under fast-growing trees to meet its fuel needs. This is still a lot of arable land for some densely populated countries of Asia, where only a total of 0.8 ha of arable land is available for every six members of the agricultural population. In urban areas the use of charcoal increases demand for fuel wood, and here too improvements could be made in charcoal production to increase the yield of charcoal from a given quantity of wood by up to four times.

The aforementioned tendency to burn manure for fuel instead of using it to enrich the soil could be overcome by the more widespread production of biogas, which allows the manure to be used for both purposes: fuel and fertilizer. The two main problems are, firstly, the need for a sufficiently high ambient temperature, particularly in the winter to keep the biogas plant running effectively and, secondly, the capital cost of the biogas plant. The Chinese have produced a cheap design which can operate with small volumes. The potential benefits of using biogas more extensively in developing countries are manifold: it would increase the availability of fuel and fertilizer, reduce the workload of those having to collect fuel wood and relieve ecological pressure from tree-felling and dung-burning. These benefits justify assistance being given to developing countries for this purpose.

ENVIRONMENTAL DEGRADATION

There is a great deal of concern about the deterioration of the environment though various causes. In order to tackle this problem effectively it is essential firstly to establish whether the grounds for concern are justified, and also the extent of the problem.

We may take the case of soil erosion as an example. It can be demonstrated that the 'gloom and doom' predictions of extensive loss of soil are exaggerated. There are data available from the USA covering several decades during which the loss of topsoil has been consistently measured (Table 1.8).

Comparisons are difficult to make because different terms were used to describe the erosion, but it would appear in general terms that the problem of erosion has diminished in the USA over the last half-century rather than deteriorated.

Table 1.8. Cropland Erosion Studies from the USDA.

Date	Finding	% of utilized agricultural land based on the sites surveyed
1934	Slight erosion	56
	Moderate erosion	32
1958	Conservation adequate	12
	Conservation needed	36
1967	Conservation adequate	36
	Conservation needed	64
1977	Erosion < 5 t (acre year)$^{-1}$	66
	Erosion 5–13.9 t (acre year)$^{-1}$	23
	Erosion > 14 t (acre year)$^{-1}$	12
1987	Erosion < 5 t (acre year)$^{-1}$	75
	Erosion > 5 t (acre year)$^{-1}$	25

Source: US Department of Agriculture (in Pimentel and Hall, 1989)

Environmentalists would maintain that any erosion at all is a bad thing and use this to suggest that agricultural production either cease or be severely limited. Whilst it may be true that it is difficult to avoid some soil erosion during agricultural production, it is also possible to improve and repair soil structure by agricultural means. It is not possible to say if there is a 'tolerable' amount of erosion but a certain amount can with knowledge and technology be repaired.

The reports of soil erosion in under-developed countries is worrying, for there the technology and funds may not be readily available to effect the necessary conservation and repair. There is an absolute choice for these countries faced with the need to grow crops for food or export, against which soil conservation is unlikely to be the higher priority. Interestingly, it is the production technique introduced by colonial settlers in many developing countries which points the way to avoiding soil erosion. The problem of erosion and leaching of minerals from deluges of rain in the equatorial lowlands can be overcome with a well-organized plantation system. This is achieved by planting an 'umbrella' of vegetation to maintain a natural barrier between the hot sun, the soil and the heavy rain. The oil palm has proved an effective substitute for the old traditional food cash crops such as cocao and bananas. In addition, the ground between the trees can be protected by shade-tolerant food crops such as legumes and yams. This combination of plants provide both food and non-food products as well as conserving the soil; they illustrate how by the application of agroforestry techniques, the twin goals of food and non-food production can be sustained.

Some of the same problems beset the question of using agrochemicals. Environmentalists complain about the deleterious effects of using agrochemicals. Unfortunately, because their use is a relatively recent phenomenon there are insufficient data to assess the accuracy of the environmentalists' claim. Serious industrial accidents involving agrochemicals, such as the release of a gaseous raw material used to make pesticides in Bhopal, India, and the agrochemical spillage into the Rhine in Switzerland, produce a sensational media response and a distorted view of the risk attached to using these tools of modern agriculture. They also overshadow efforts to make agrochemicals safer for use. Environmental Protection Agencies throughout the world are responsible for enforcing a ban on agrochemicals considered unacceptably dangerous to use. In addition they have encouraged the manufacturers of agrochemicals to produce chemicals which are

- more target-specific;
- more effective at lower concentrations;
- readily biodegradable.

Biotechnology offers the opportunity to reduce the volume of agrochemicals used by breeding in resistance to disease and pests. This is the basis of a strong argument to oppose the anti-technology lobby which seeks reversion to old low-technology methods as a means of preserving the environment. The judicious use of agrochemicals can produce more crops for either food or non-food use with no significant damage to the environment. There is also growing interest amongst farmers in new methods of agrochemical use which not only protect the environment but help improve profitability. There is a good example of this from the more accurate dosing of nitrogen fertilizer for sugarbeet production. A soil test prior to drilling will tell the farmer just how much nitrogen should be applied. Accurate applications of fertilizer actually help to raise the value of the crop whereas overdosing would depress the sugar content.

A third objection to the extension of modern agricultural farming methods to the production of crops for non-food purposes is the risk of losing biological diversity. There is a fear that the production of the main commercial crops for biomass, perhaps in areas previously not used for agricultural production, will lead to a loss of species. In fact, a lot depends on the character of the land before it is brought into use. Bringing degraded land into use for biomass production may actually help to maintain the diversity of species. It also depends on what types of plants and techniques are used to produce biomass. There are objections to the technique of cloning, which is used to propagate genetically identical plants from the cells of leaf cuttings. While cloned plants may reduce biodiversity if planted as a monoculture, farmers can also grow clones of each of several species on the same site. The whole point of using clones is that they have been genetically selected for a supposed superior genetic trait. They can therefore increase the capacity of the biomass to flourish in an unpredictable environment.

Biomass energy can be produced from a wide range of plant residues and agricultural wastes, making a positive contribution to the environment in the process. This range is illustrated in Table 1.9. Not all of this energy is recoverable, but it has been estimated that of the annual 12 exajoules (1 exajoule = 10^{18}J), at least 6.7 could be recovered. Wood mill residues have historically caused considerable environmental damage, such as the poisoning of lakes in Scandinavia, and a self-contained use for these residues would thus be a welcome solution for the industries that produce them.

There are several other examples of ways in which a marriage can be made between the objectives of sustainable biomass production and maintaining biodiversity. One of these is interplanting with nitrogen-fixing plant species. They not only reduce the need for artificial nitrogen fertilizer (and thus reduce the risk of water pollution) but also

Table 1.9. Biomass residue resources in the USA/exajoules year^{-1}.

Forest residues	1.84
Wood mill residues	2.30
Maize	2.27
Wheat	1.21
Soyabeans	1.03
Sorghum	0.25
Barley	0.22
Rice	0.11
Oats	0.14
Rye	0.01
Sugarcane	0.22
Cotton	0.06
Orchard residues	0.12
Manure	0.29
Urban refuse	1.39
Urban wood waste	0.60
Total	12.06

Source: National Audubon Society (1991)

reduce the risk of crop failure and increase the diversity of species compared to monoculture. Another example is the use of 'time-released' nutrients which match the use of nutrients to the need for them over time. This reduces costs and waste, as well as the risk of contamination of water supplies from over-abundant use of agro-chemicals.

The production of biomass for energy can make a positive contribution to environmental protection. There are political efforts to reduce the amount of carbon dioxide emissions which are considered the main cause of the greenhouse effect. The USA has declared a willingness to halve its carbon dioxide emissions from the use of fossil fuels. To replace the equivalent amount of fossil fuels with bioenergy it would be necessary to plant 30 million ha with renewable biomass. By coincidence this corresponds to the area currently withdrawn from food production under the Acreage Reduction Program, under which growers are paid not to produce. Not all of the land withdrawn is suitable for biomass production, but it is likely that land surplus to food production requirements is likely to increase in future. Similarly, in Europe large areas of agricultural land are being taken out of food production, partly in response to internal pressure to reduce the cost of disposing of food surpluses and partly in response to external pressure through the General Agreement on Tariffs and Trade (GATT) to reduce barriers to trade. It is not practical for agriculture to replace fossil fuels completely with biomass energy, but a significant contribution can be made.

INTERDEPENDENCE IN AGRICULTURAL PRODUCTION

The key to the optimum use of the Earth's resources for food and non-food production is greater interdependence in agricultural production. This direction for agriculture in the future has little historical precedent. Indeed, historically nations have insisted upon self-sufficiency in agricultural production, born of a suspicion of the strategic risks involved in depending upon other nations for vital agricultural raw materials. There are comparatively modern examples of how these fears are worked out even between economically interdependent nation states. The European Economic Community was established by the Treaty of Rome in 1957 on the basis that economic co-operation and integration were the best security against a repetition of the conflicts of two world wars. A Common Agricultural Policy was developed to ensure security of food supply, but in practice it still falls a long way short of optimum economic interdependence between member states. Instead of reliance on supplies of agricultural produce from countries better placed to produce beyond their own needs at lower cost, there is inefficient duplication. The comparative advantages of the North European states to produce cereals and of the Southern European states to produce oil crops are not fully exploited. The effect of governments insisting upon self-sufficiency even in a common market of nation states has been to raise the support price of agricultural commodities to the level of the less efficient, and thus to distort market prices and encourage over-production.

This failure in miniature is greatly magnified at global level, where such unwillingness and incapacity to interdepend lead to a serious shortage of agricultural raw materials in many countries. The obsession with self-sufficiency has led to the gross distortion of world agricultural markets that the GATT negotiations have tried to address at the time of writing. If only governments would accept the notion of comparative advantage in agriculture, a more rational world agricultural policy could be achieved. In Ricardo's system of comparative advantage, nations would specialize according to the comparative advantage amongst all goods and services and the economic output of the whole world would be greater. Comparative advantage does work to a limited degree in agriculture. For example, the USA has an absolute advantage in being able to produce wheat more cheaply than its competitors. But other countries do manage to export wheat where they have a comparative advantage, i.e. where they can grow and export wheat better than most other goods. If only the criterion of food self-sufficiency could be removed, the potential for increased production, greater diversity and cheaper agricultural production would be increased because certain countries have very considerable absolute advantages which are under-utilized.

The commodity-by-commodity insistence upon self-sufficiency aggravates the problem of world hunger. To cure hunger would mean supplying 50 million people with 6.5 bushels (1 bushel≃35 l) of grain each, a total of 300 million bushels; this is only one tenth of the present world surplus in grain-exporting countries. To cure malnutrition 900 bushels would be required, but this is still small compared to the surplus and takes no account of the land retired from food production or future improvements in yield. There is thus no justifiable reason why the Earth's resources could not support up to four times the present population size on a reasonable diet, as well as providing renewable energy, if only agricultural policies could be inspired by true economic efficiency and a willingness to interdepend for agricultural production.

Conclusion

This chapter has looked at the way that agriculture and land use has developed over time. Humans have diversified their resources and applied new technology continuously to raise yields and improve the efficiency of agricultural production. The modern world now faces a series of crises, including the projected exhaustion of fossil energy. History shows that by the application of new technology and diversification such impasses can be overcome. This chapter has argued that the world has sufficient resources to cope with the increased demand for food from a growing population as well as the capacity to meet demand for some non-food production, and biomass energy in particular. By bringing degraded or neglected land into production it is possible to tap some of the Earth's unused land resources for these purposes. A sustainable system of production can be achieved by the use of the new techniques of biotechnology coupled with more judicious use of the resources presently available.

References

Bews, J.W. (1973) *Human Ecology*. Russell Russell, New York, 312 pp.
Calvin, M. (1986) Letter to D. Pimentel 18.12.86.
Clark, C. and Haswell, M. (1970) *The Economics of Subsistence Agriculture*. Macmillan, London, 267 pp.

Cotterel, F. (1955) *Energy and Society*. Greenwood Press, Westport, Connecticut, 330 pp.
Food and Agriculture Organization (1981) *Agriculture: Towards the Year 2000*. FAO, Rome.
Forbes, R.J. (1968) *The Conquest of Nature*. Frederick Praeger, New York, 98 pp.
Hall, D.O., Barnard, G.W. and Moss, P.A. (1982) *Biomass for Energy in the Developing Countries*. Pergamon Press, Oxford.
Jennings, J.S. (1989) *The Energy Outlook – its Implications for Oil and Gas*. Energy Policy Seminar, Sanderstolen, Norway 9.2.89, Shell International Petroleum Company Ltd.
Leach, G. (1976) *Energy and Food Production*. IPC Science and Technology Press Ltd, Guildford, Surrey, 197 pp.
Lee, R.B. and DeVore, I. (eds) (1976) *Kalahari Hunter–Gatherers*. Harvard University Press, Cambridge, Massachusetts, 160 pp.
Lowry, J.H. (1986) *World Population and Food Supply*. Edward Arnold, London.
Meadows, D.H. and Meadows, D.L. (1992) *Beyond the Limits*. Earthscan Publications Ltd, London.
National Audubon Society (1991) *Toward Ecological Guidelines for Large Scale Biomass Energy Development*. Executive summary of a workshop held on 6.5.91 at Princeton University, Princeton, New Jersey.
Pimentel, D., Terhune, E.C., Dyson-Hudson, R., Rocherean, S., Samis, R., Smith, E.A., Denman, D., Reifschneider, D. and Shepherd, M. (1976) Land Degradation: effects on food and energy resources. *Science* 4, 149–55.
Pimentel, D. and Pimentel, M. (1979) *Food, Energy and Society*. Edward Arnold, London.
Pimentel, D. and Hall, C. (1989) *Food and Natural Resources*. Academic Press, San Diego, California.
University of California (1977) *Water*. California Agriculture, Berkeley.
Wall Street Journal (1992) *Proposed carbon tax is controversial at the Earth Summit in Rio*. European edition, 9.6.92.
White, L.A. (1943) Energy and the evolution of culture. *American Anthropology* 45, 335–354.
World Resources Institute (1990) *World Resources 1988–9*. World Resources Institute and United Nations, New York.

2

WHICH RAW MATERIALS WILL BE USED?

Not only have there been shifts in the pattern of uses of agricultural raw materials over the centuries, but the range of crops and animal products has changed as different attributes were sought from the produce of the land. From a global viewpoint, the range of crops which can be used for non-food purposes is very large but in practice only a small amount of this potential is used on a commercial basis; this is due to technical, economic and climatic constraints. New technology is being applied to remove these constraints, for example, to allow crops previously confined to tropical zones to be grown in temperate areas. The end-products sought for non-food use are principally carbohydrates, fats and fibres. This chapter will explore not only the conventional crops used to produce these products, such as wheat, maize, sugarbeet, rape, sunflower, flax, hemp and cotton, but also less conventional crops such as lupin and elephant glass, and agricultural by-products such as short-rotation woods, forestry products, straw and agricultural waste.

CARBOHYDRATES

The vigour and importance of the chemical industry has been based largely on the chemistry of carbon, first from coal, then from oil, and now increasingly from agricultural sources. The most common sources of carbon used are sugars, such as glucose and fructose, and starch. These carbohydrates occur naturally in a wide range of plants which cover all the climatic zones of the world. Conventional sources of carbohydrate include a range of cereals such as wheat, rice and maize, as well as the commercial sugar crops, beet and cane. Less conventional sources of carbohydrate which have yet to be fully exploited commercially include sorghum, Jerusalem artichoke and sweet potatoes or yams.

There is abundant carbon available in cellulose, but it is not yet commercially viable to recover this polysaccharide from otherwise cheap sources such as straw. This is a good example of where a technical breakthrough could change the whole economic picture of carbohydrate use. If cellulose could be hydrolysed more readily it could be a very cheap source of glucose and fructose syrups.

Sugar

Sugar is the common term used for sucrose ($C_{12}H_{22}O_{11}$); its chemical structure is shown in Fig. 2.1. The term 'sugar' is also used more loosely to refer to all water-soluble, mono-, di- and oligosaccharides. Examples of these are glucose and fructose (monosaccharides), and sucrose, lactose and maltose (disaccharides). Sucrose occurs naturally in a wide range of plants but it is accumulated to high enough concentrations for commercial exploitation in only a few crops, the best-known of which are sugarbeet and sugarcane (in cane the average sugar content is 13% and in beet 16%). Sugarcane was the first to be developed commercially and since its natural habitat is tropical, the product had to be imported by countries in temperate zones unable to grow it. It was the naval blockade of continental Europe during the Napoleonic wars which led to the discovery that a Mediterranean weed beet could be used to produce sugar. Modern beet varieties have been bred to increase their sugar content and are now widely grown in temperate climates throughout the world. The sugar content of beet is typically 16%, which means that from 1 t of clean beet 160 kg of sugar can theoretically be recovered. In fact part of that sugar resides in the molasses – the residue left after extraction and crystallization of the white sugar. Molasses has a sucrose content of up to 75%.

Sucrose has many advantages as a raw material (Kollonitsch, 1970). It is a well-defined pure chemical, soluble, and easily handled in aqueous solution. It is stable and can be stored either as a liquid or a solid. It can be utilized in different degrees of purity by various sections

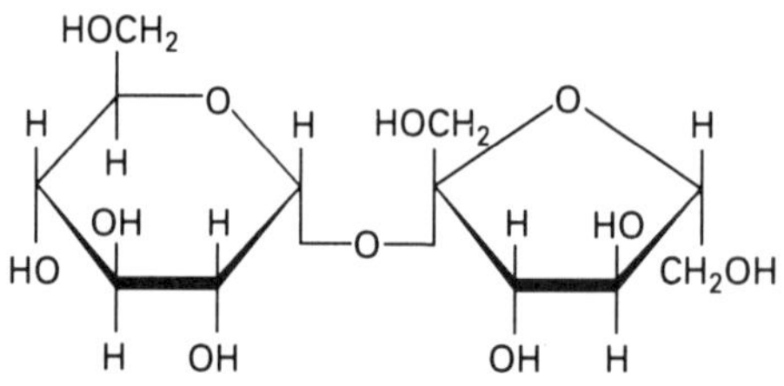

Fig. 2.1 Structure of sucrose.

of the chemical industry (Pennington and Baker, 1990). Sucrose in its least refined form can be purchased in molasses which is suitable for fermentation but gives lower productivity and, possibly, effluent problems. Purer forms of sucrose usually cost more, but give higher productivity and less environmental problems. Pure sucrose is used in traditional chemical reactions (Lichtenhalter, 1990) to synthesize sugar derivatives used as intermediates in the chemical industry. Sugar ethers, once produced by reaction with organic sulfates and organic halogen derivatives, are now made by newer methods that include the preparation of sodium 'sucrates' and their subsequent reaction with organic halide to produce the desired ether. The potential applications of sucrose ethers include surfactants and surface coatings. Sugar esters, including mono- and poly-substituted esters of aliphatic acids like acetic and propionic acid, as well as the longer-chain saturated and unsaturated fatty acids, are used principally in the manufacture of detergents, and also as emulsifiers in cosmetic applications.

Miscellaneous reactions of sucrose include its reactions with aldehydes and ketones and with alkali metals, together with pyrolysis, oxidation, hydrogenation, reductive aminolysis and halogenation.

The fermentation of sucrose by various microorganisms (Dordick, 1991) is used to make a range of products that includes alcohol (ethanol), various solvents, carboxylic and amino acids, polysaccharides, enzymes, antibiotics, vitamins, steroids, biodegradable plastics and single cell protein.

Enzymatic processes based upon sugar can produce or convert proteins and peptides, carboxylic and amino acids, oligo- and poly-carbohydrates with special functions, lipid substitutes, and also nucleic acids, nucleotides and their derivatives.

Many applications of sucrose depend less on its chemical reactivity than on its physical properties. The pharmaceutical industry uses sucrose in several different forms on account of its sweet taste, its crystalline structure and its bulking properties. Tablets and other pharmaceutical preparations are frequently coated with sucrose-containing glazes to give a protective finish to the product. Sucrose is also used as a diluent or base for preparing encapsulated or sustained-release-dosage pharmaceuticals. Sugar is also a key ingredient in a simple remedy for diarrhoea: when mixed with boiled water and a little salt it helps to replace lost body fluids. Dehydration is a critical condition particularly for children and this inexpensive remedy has proved important in developing countries. Non-food uses of sugar remain small compared to their food uses, but the latter ensure a sure supply for the former. The major non-food uses of sugar likely to expand are those depending on biochemical procedures and the use of enzymes.

Starch

Starch is widely present in almost all plants. It is abundant in cereal grains and tubers and is also present in peas and beans. The most commonly consumed sources of starch are maize, wheat, potatoes and rice. These are grown for both food and non-food uses of starch. Other commercial crops grown for their starch content include cassava, manioc and sorghum, but their use tends to be exclusively in the food sector and they have not been developed for non-food uses.

Chemically, starch is a mixture of two homopolymers of α-D-glucopyranoside: amylose and amylopectin (see Figs 2.2 and 2.3). Most cereal starches are made up of about 75% amylopectin and 25% amylose molecules, but root starches have a higher amylopectin content and in waxy corn starches it may be as high as 100%. At the other extreme, high-amylose corn starch and wrinkled pea starches contain 60–80% amylose. The chemical structure of starch is much more complicated than that of sucrose. It is not known what properties in starch make it work in a particular way. Granules of starch are not soluble in cold

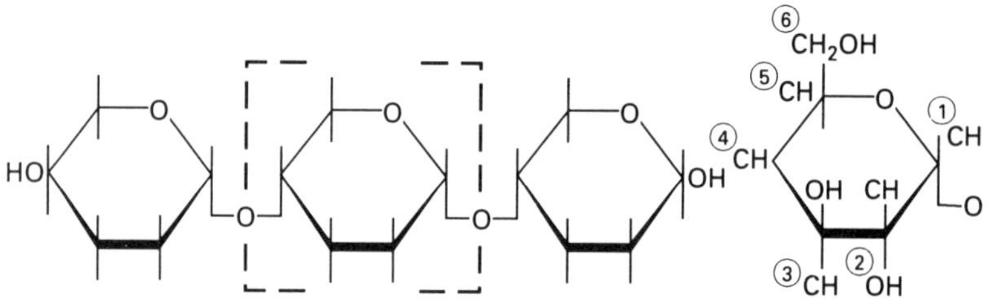

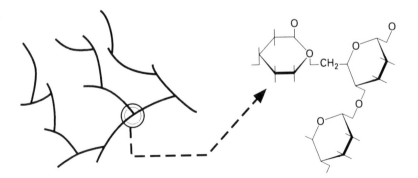

Fig. 2.2 The structure of the linear fraction (amylose) of starch, linkage α(1–4).

Fig 2.3 The structure of the branched fraction (amylopectin) of starch, branching linkage α(1–6).

water but they absorb water and swell rapidly when heated to a gelatinization temperature range. When allowed to cool, the hydrated molecules begin to precipitate, forming a rigid gel. These properties afford starch a wide range of food and non-food applications.

The methods of extracting the different types of starch vary according to the attributes of the raw material.

Maize starch extraction

The most common process for recovering starch from maize is a wet milling process. There are four separate phases in this process.

Steeping of the grains
The maize grains are placed in steeping tanks of water to make them swell and soften in order to facilitate the separation of the different parts of the grain, i.e. the starch, the germ, the hull and the soluble products. The liquid left after this steeping process is known as 'corn-steep' and can be sold on as a feedstock for the microorganisms used in antibiotic manufacture.

Germ separation
The swollen and softened grains are then milled to free the germ, which is then washed, dried and pressed to extract maize germ oil. The residual germ cake is sold as an animal feed.

Milling and sieving
The remaining liquid contains starch, protein and fibre in suspension. It is then finely sieved to separate out the different elements.

Separation of starch and gluten
Centrifugation is used to separate the gluten (a protein) from the starch. The starch slurry is then purified. Typical yields of these different products recovered from maize in the process of starch manufacture are as follows:

100 kg of maize yields   62–63 kg starch
                          3 kg crude maize oil
                        19–20 kg maize gluten feed
                          4–5 kg maize flour
                          4 kg maize germ cake.

Wheat starch extraction

The most commonly used cereal for starch production in Europe is wheat, although maize starch remains the most commonly used worldwide. The reason for this is that a high-value co-product, known as vital wheat gluten, can be obtained. According to the quality of the wheat a wide range of figures for starch extraction from 100 kg of raw material exist: the yield from 100 kg of wheat falls with the ranges 49–52 kg starch with 6–7 kg gluten. The remainder consists of protein effluent, syrup residues and wheatfeed. The process for recovering wheat starch therefore has to avoid damaging this product, and certain chemicals used in other starch extraction processes therefore cannot be used for wheat. Depending on the process used to extract starch, the gluten can vary in vitality; 'vital' gluten refers to good quality.

The processing methods have taken the technology from the maize wet milling process and adapted it to the treatment of wheat. One of the most commonly used commercial methods is known as the Martin method. Wheat flour is used as the raw material on account of the technical difficulties of using the whole grain directly. The flour is mixed into a dough with water and rested to allow the gluten to hydrate. It is then fed into an extractor vessel where the starch is washed away from the gluten. Care is required to prevent the gluten from being dispersed or broken up, and to allow it to remain a single coherent mass. The remaining starch slurry is further purified by sieving and screening. It is then centrifuged to extract the water and the remaining starch is dried in a flash drier. The effluent is either disposed of as waste, or dried for cattle feed, or used as a nutrient for antibiotic manufacture. In terms of value, the major product of this process is the vital wheat gluten.

In another common process for recovering gluten and starch from wheat, known as the Batter process, the wheat flour dough is dispersed in water and the gluten is recovered in a sieve. The dough gradually dissolves leaving small curds of gluten in a starch slurry. Water is then extracted from both products and they are dried.

A third, less commonly used method involves using an alkali to separate the protein from the starch. Wheat flour is mixed with dilute caustic soda solution, the resulting liquor is screened and the starch and dispersed protein separated through a centrifuge or filter. This process is repeated using dilute caustic soda on the solid starch cake but the gluten produced is usually of poor quality. An alternative and better reagent for use in this process is ammonium hydroxide.

The starch extracted by these methods can be sold on for industrial use in either dry or liquid form. The starch may be graded according to the size of the starch granule, for example, as A grade (the purer fraction, containing larger starch granules) or B grade (less pure, smaller or damaged granules) starch.

More recently attention has turned to enzymic methods of recovering starch, and the large European enzyme manufacturer Novo is at the forefront of these developments. However, for starch manufacture from maize, enzymic methods are not usually seriously considered on account of their cost.

Potato starch processing

Potato starch is known as farina. There are specific varieties of starch potatoes and their attraction to industrial users is that they can be used to obtain a finer starch powder than that obtained from wheat or maize. Starch potatoes tend to be grown in limited areas of northern Europe; for example, in a certain area of the Netherlands they are the main crop and they enjoy a premium price. Unlike wheat and maize, potatoes cannot be stored for long periods of time on account of the amount of water they contain. Consequently, starch extraction usually takes place during the winter months following autumn harvesting. The potatoes are usually delivered with soil adhering to them, and with loose soil and stones in the load which have to be removed. So the first step involves washing, and a certain amount of starch will leach out from broken or damaged potatoes in this process. Once clean, the potatoes are pulped using a rasping machine and a hammer mill. The pulp is then separated into fibre, starch slurry and proteins by conventional screening methods. The starch slurry is purified by further washing, then dehydrating and drying.

100 kg of potatoes yields about   15–23 kg starch
　　　　　　　　　　　　　　　　　2 kg pulp
　　　　　　　　　　　　　　　　　1 kg protein
　　　　　　　　　　　　　　　　　75 kg water

Rice starch processing

As in the treatment of maize, rice grains are steeped, usually in dilute caustic soda solution, to cause them to soften and swell before use. This allows the starch granules to be separated out. The wet rice grains are then ground in a mill to a mash. This mash is then screened and the starch slurry separated from the protein in a centrifuge. The starch is further purified and concentrated before de-watering and drying.

100 kg of rice yields about   65 kg of rice starch
　　　　　　　　　　　　　　　5 kg fibre
　　　　　　　　　　　　　　　7 kg protein
　　　　　　　　　　　　　　　12 kg waste products

Table 2.1. Differences between commercial native starches.

Source	Wheat	Maize	Potato	Rice
Granule diameter/mm	10	15	30	20
Moisture/%	13	13	19	13
Lipids/%	0.8	0.6	0.05	0.1
Paste viscosity	high	medium	v. high	high

Starch derived from these different sources is known as 'native' starch and has usually been dried to a moisture content of around 13% (19% for potato starch). The properties of starches from different sources differ and have been summarized in Table 2.1, taken from an unpublished survey undertaken for the EC by the Centre for European Agricultural Studies (CEAS) and the Institut de Gestion Internationale Alimentaire.

The different attributes of these native starches help to determine their use. So, for example, paste viscosity is important for starch used in the adhesive and paperboard industry.

From native starch, a wide range of derivative products can be obtained, from glucose syrups to alcohols to esters and ethers of starch. An indication of the uses of some of these derivatives is given below.

- Sorbitol is a sugar alcohol obtained by subjecting a starch hydrolysate to a process of hydrogenation, purification and concentration. Its non-food uses are principally in the pharmaceutical industry as a moisture stabilizer and anti-crystallization agent, e.g. in toothpaste.
- Dextrins are produced by subjecting dry starch to heat treatment in the presence of catalysts. They are used chiefly in the paper and textile industries and as fat substitutes in food.
- Starch esters are used as stabilizers for paints, oils and cosmetics, and to produce a clear film which can be made soluble or non-soluble for the food industry, and the textile and paper-sizing industries.
- Starch ethers can be used as protective coatings for wood, metal and glass, or as adhesives such as printing and poster pastes.

Starch is thus an extremely important and versatile agricultural raw material for non-food use. In Europe the main non-food application of starch in volume terms is in the paper and carton industry, followed by plastics and resins, organic chemicals, enzymes and glues, and pharmaceuticals (see Fig. 2.4).

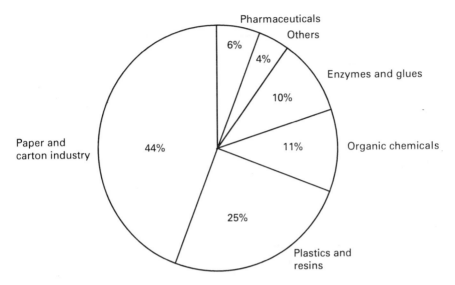

Fig. 2.4 Non-food uses of starch in Europe.

Straw

As a by-product, straw offers a relatively cheap source of carbohydrate. Annually in the UK alone, 6–7 million tonnes of straw have to be burned or incorporated into the soil for lack of alternative use. From 1993, straw will no longer be allowed to be burned. Cereal straw contains 70% carbohydrates, primarily cellulose and pentosans. Pentosans can be hydrolysed to sugars, mostly glucose and xylose; however, the process of hydrolysis has not yet been developed to the point where it is economically viable. Cellulose is extremely resistant to hydrolysis, however, because it has a highly ordered crystalline structure and because lignin present acts as a physical barrier. The most popular approaches to converting cellulose into glucose are dilute acid hydrolysis, concentrated acid hydrolysis and enzymic hydrolysis.

Dilute acid processes give glucose yields of about 50% but the lignin is also seriously degraded by the process. The use of concentrated sulfuric or hydrochloric acid can achieve higher glucose conversion rates (85–90%) but the material costs are higher and the lignin is adversely affected. Enzyme processes yield pure products and do not damage the lignin but the glucose conversion rates are low (in the region of 50–60%), the reaction takes a long time and it is usually necessary to pretreat the feedstock. The pretreatment of straw usually involves a physical process of breaking down the raw material using heat (as steam) and pressure. The use of hydrogen fluoride offers the

Table 2.2. The attributes of softwood and straw.

	Softwood	Straw
Cost	£55 t^{-1}	£30 t^{-1}
Moisture content/% of dry matter	140	16
Bark	10%	none
Average cell size/mm	2.7	< 1.5
Fibre length/mm	1.0	0.64
Waxes/mm	none	> 0.5

Source: Bolton (1992)

prospect of minimal feedstock pretreatment, high rates of glucose conversion (90%) and undamaged lignin. The disadvantages of this process are that hydrogen fluoride is relatively costly, dangerously corrosive and highly toxic. Another process employing phenol as solvent allows part of the lignin and part of the hemicelluloses to be extracted before hydrolysis of the cellulose. The other main components of straw can then be separated in a relatively pure form. The process, known as solvolysis, can be carried out at atmospheric pressure and 100°C.

Another major opportunity for straw use is as a fibre base for paper and particle boards. This is a particularly attractive option in Europe, where most countries have to import wood and wood pulp to meet their needs. Globally, in the context of concern about the destruction of forests, the use of straw as an alternative to wood offers environmental advantages.

Table 2.2 shows a comparison of the attributes of straw with those of softwoods. One of the most significant contrasts here is the moisture content; obviously, a lower moisture content leads to a saving in processing costs. The waxes contained in straw offer an environmentally friendly alternative to hydrocarbon waxes. When one considers that softwoods are produced on approximately a 40-year cycle compared to the annual production of straw, these advantages are highlighted. Straw contains many other components than just cellulose, and so fractionating the straw crop is an essential part of optimizing the use of every part of the plant. The leaf of straw is best used as an animal feed and the stem can be further fractionated for different industrial uses. The best method for doing this has been found to be thermomechanical pulping. In the UK, funds have been made available under the government-sponsored LINK programme for precompetitive research on straw. The European Commission has made research funds available for the upgrading of straw into paper and for fermentation into ethanol under the OPLIGE project for the optimization of lignin crops through genetic engineering.

Sweet sorghum

Sweet sorghum is a cereal grass that is widely grown in Africa, Asia and the Americas. In Europe, the production of sweet sorghum is confined for climatic reasons to southern Europe. Here it has demonstrated excellent productivity and efficiency of water usage by producing 8.2 kg of dry matter per litre of water consumed. The plant is still used principally for fodder but could be used in future as a carbohydrate substrate for fermentation; it can also be a source of fibre.

Jerusalem artichoke

This is a carbohydrate source which has been rediscovered as a potential raw material for non-food uses. For example, when used for alcohol production it can yield more ethanol per hectare than conventional crops used in alcohol production, although the storage compound in the Jerusalem artichoke is inulin, not starch. It can be grown widely all over the world. The main obstacle to its use is the difficulty of harvesting a small irregularly shaped tuber. Furthermore it has proved even more difficult to store than the potato. The tuber can be used for the production of pure fructose (inulin can be hydrolysed to fructose) or the total plant can serve as a source of biomass.

OILS AND FATS

The chemical industry uses both animal and vegetable fats for non-food purposes (Shields, 1992). Worldwide, 10.5 Mt (1 Mt = 1 million tonnes) of oils and fats are used by the chemical industry, that is about one third of total world oils and fats consumption. The EC chemical industry alone uses 3 Mt of these raw materials annually; of these, 1.3 Mt are from inedible sources such as tallow, linseed and castor and 1.7 Mt from edible sources such as palm, coconut and soya. For cost reasons, the chemical industry often tries to use lower-quality edible fats or the waste products of food oil processing. So there is a complementary use between the food and non-food uses of oils and fats. The prices of oils and fats vary considerably; Table 2.3 indicates the range.

Oil-bearing plants have traditionally been bred to produce high yields of oils suitable for human consumption (except for toxic oils like linseed and castor). The palatability, texture and physical form of the oils are important for food use but of minor significance to the chemical

Table 2.3. Current market prices for selected oils and fats/US$ t^{-1}.

Rape oil	383
High-erucic rape oil	655
Coconut oil	437
Tallow	355
Palm oil	366

Source: Shields (1992)

industry, where the chemical composition of the oil is the most important factor. It is the fatty acid fraction that interests the chemical industry:

- the chain length of the fatty acids – narrower distributions of the chain length, and short and medium chain lengths, are desirable;
- double bonds – number, position and structure of double bonds in greater clarity;
- existence of interesting functional groups, e.g. hydroxy- and epoxy-groups.

When the chemical industry wishes to use agricultural raw materials it usually has to work with a plant which has been bred to meet the requirements of the food industry. A good example of this is the emphasis placed upon reducing the erucic acid content of oilseed rape, which is not acceptable for food use. However, it was precisely the varieties containing high levels of erucic acid that the chemical industry wanted. High-erucic rape varieties are a potential source of greases and lubricants. The world market for industrial lubricants is estimated to be 1.9 million tonnes. The main advantage of vegetable oils over mineral oils is that they are better for the environment on account of their biodegradability. One potential use could be in oil-based drilling fluids. At present mineral oils are incorporated in drilling mud used in the extraction of oil, but this has been found to damage the environment within a radius of at least 2000 m of the drilling site. At present rapeseed oil is about six times as expensive as the mineral oil, but government-funded research aided by commercial sponsors is being undertaken with the aim of reducing the cost.

Both high- and low-erucic rapeseed varieties are available and continue to be developed to serve both the food and non-food markets. The non-food processor often requires the use of expensive separation and purification techniques, which inevitably produce lower-value by-products for which uses must be found. Furthermore, the chemical industry is more exposed than the food industry to the vagaries of world market price movements of the oils they use, because whereas the food industry can switch from one oil feedstock to another, the chemical industry is dependent on the specific chemical composition of the oils.

This dependency of industrial users on the food industry, its by-products and its inferior feedstock is likely to remain in future unless the consumption of oils and fats for non-food purposes increases to an extent which justifies tailor-made crops being bred and grown. The low volume is not the only disincentive to the production of crops bred specifically for industrial (i.e. non-food) use; another problem is that many of the oil-bearing plants are grown in tropical zones and efforts to adapt these plants to grow in temperate climates have been only partially successful. The chemical industries of the developed world are mostly located in the Northern Hemisphere, and certainly for pilot schemes for non-food uses of oils they tend to import the oils. An idea of the extent to which Europe depends on imported oils can be gained from Table 2.4.

The European chemical industry uses oils and fats derived from crops to produce a wide range of products which are consumed by several important non-food industries. The approximate oil equivalent consumption in each of the major areas is given in Table 2.5.

The uses of oleochemical products cover a wide range. They are discussed by categories below.

Table 2.4. European marine and vegetable oils and fats balance, 1990/ × 10^3 t.

Oil or fat	Production	Import	Export	Balance	Total/%
Olive oil	1194	79	154	1119	12.06
Groundnut oil	4	201	4	212	2.28
Soya oil	2335	10	706	1614	17.39
Colza/rapeseed oil	2355	117	939	1294	13.94
Sunflower oil	1653	177	293	1549	16.69
Sesame oil	–	2	–	2	0.02
Cotton oil	70	4	1	73	0.79
Coconut oil	57	543	5	590	6.36
Palm kernel oil	12	408	6	418	4.50
Linseed oil	62	46	16	93	1.00
Caster oil	20	87	2	107	1.15
Corn oil	152	79	32	188	2.03
Palm oil	–	1319	20	1315	14.17
Others	39	59	18	145	1.56
Total	7953	3131	2196	8719	93.94
Marine oils	111	480	28	563	6.07
Grand total	8064	3611	2224	9282	100.0%

Source: FEDIOL (the European Federation of Oilseed Crushers and Processors), annual conference (1991)

Table 2.5. Consumption of oils and fats by non-food industries/$\times 10^3$ t oil equivalent.

Alkyds	100
Oleochemicals	900
Soap	400
Animal feeds	600
Others	1000
Total	3000

Source: European Council of Chemical Manufacturers Federations (CEFIC)

Detergent industry

Detergents are the largest single end-use of oleochemicals. The uses of vegetable-based surfactants are growing at the expense of petrochemical-based detergents on account of their environmentally preferable properties. Examples of these end-uses are washing substances, emulsifiers, dispersants, anti-foaming agents, softeners, viscosity regulators, superfatting agents and detergency improvers.

Paints industry

The paints industry uses oleochemicals in the manufacture of alkyd resins, dispersing agents, suspending agents, thixotropic agents, defoamers, pigment dispersers, air release agents, levelling agents, epoxy plasticizers and special plasticizers.

Lubricant oil additive industry

This is the third largest market for oleochemicals, and the global market is equivalent to 2.9 Mt annually. This is likely to expand further at the expense of petrochemicals, for example, with the insistence on the use of biodegradable lubricants as marine oils in northern Europe. Other examples of oleochemical uses in lubricants are as emulsifiers, anti-corrosion agents, anti-foaming agents, thickeners, viscosity index improvers and extreme pressure additives.

Plastics additives industry

Lubricants, antioxidants, plasticizers, stabilizers, dilution agents for pastes and complex esters are all produced from oleochemicals.

Pharmaceutical industry

Here the relevant end-uses are in ointment bases, suppository bases, oily components, emulsifiers, tableting auxiliaries and consistency factors.

Cosmetics industry

The end-uses include cream bases, emulsifiers, consistency factors, oily components, washing active substances for shampoos and for foam baths and shower gels.

Technochemical industry

Oleochemicals are used in the manufacture of a vast range of additives for the textile, leather, paper, construction, petroleum and ore flotation industries. These include dispersants, emulsifiers, wetting agents, low-foaming wetting agents, fatting agents, oiling agents, spooling oils, foaming agents, anti-foaming agents, flotation agents, aldehydes, thickeners, humectants, foam-stabilizers, biocides, de-inking agents, algicides, softeners, air-entraining agents and inhibitors.

Oils and Fats Chemical Research

Fats (lipids) from plants can offer alternative, renewable materials for chemical specialities. Numerous minor seed oils, as well as animal and some bacterial fats, are potential renewable sources of lipids from which oleochemicals could be produced. A revival of interest in the isolation, characterization and analysis of these materials using the most modern techniques could point the way towards new sources of raw materials. Unusual fatty acids containing epoxy-, hydroxy- and keto-substituted, conjugated, olefinic and branched chains are known to exist in nature.

At present the appropriate plant species have not been developed as sources of industrial raw materials, but have been identified as having potential which could be developed by a programme of conventional plant breeding and the application of biotechnology. The main aim of breeding will be to obtain unusual fatty acids, such as petroselinic acid from coriander and calendic acid from *Calendula*. Some other novel plant species of interest are discussed below.

Jojoba

These perennial shrubs grown in semiarid areas of the USA, Australia and southern Europe produce a liquid wax ester of an unsaturated

C_{20-22} alcohol with an unsaturated C_{20-22} acid. Jojoba oil is already used extensively by the cosmetics industry, and also in the USA as a lubricant.

Crambe
These are annual plants of tropical origin which produce an oil high in C_{22} monosaturated fatty acid. The plant has a 50% oil content containing 50% erucic acid. The subsequent meal which can be recovered is rich in glucosinolates – anti-feeding factors that reduce the value of the meal as a protein source.

Cuphea
These annual plants of tropical or subtropical origin produce short-chain (C_{8-10}) or medium-chain (C_{10-14}) fatty acids. It has proved difficult to domesticate this plant as a crop species in Europe and the United States; for example, there have been problems with seed shattering.

Sunflower
Sunflowers are well-known and widespread annual plants which may be capable of improvement to yield 90% oleic acid compared to 50–60% oleic acid at present.

Euphorbia lathyris
The seeds of this perennial plant of tropical semiarid regions have a 50% oil content, comprising 60% vernolic acid and 12–13% epoxy acids. The disadvantage of the plant is that the oil-bearing kernels tend to fall off before they can be harvested.

Chinese Tallow Tree
This tree produces very high yields per hectare of two oils: tallow oil and stillingia oil.

Other oil-bearing plants include members of the genus *Lequerella*, annual or perennial herbs with seed oils that are rich in C_{20} monosaturated hydroxy acids. *Vernonia* and *Stokesia* are annual plants which are of interest because their seed oils are rich in epoxyoleic acid, while those of *Limnanthes* (meadowfoam) contain 75% C_{20} fatty acids.

Basic research was largely neglected in the 1950s as the petrochemical industry expanded. Previously there had been a large number of institutes and professorships dedicated to basic research into oleochemistry. Current oleochemical research is mainly devoted towards applications research, technology improvement and environmental control. The European Commission has established a research project for vegetable oils for innovation in the chemical industry (VOICI) and for seed oils for new technical applications (SONTA), for which a budget of 80M ecu over five years (1988–1993) has been made available. Its objectives are to optimize harvesting and growing techniques of less conventional oil crops and to work on the breeding, storage, processing and end-use of these oil crops. There are, however, some constraints on the use of Commission research money: the research has to be at a pre-commercial stage and more than one member state has to be involved in the research. These have not in practice proved to be severe constraints.

Recent developments in biotechnology have opened up many possibilities for the genetic engineering of oilseed crops to produce improved plants. Progress is most likely to come from transgenic plants making 'oils-to-order' – for example, by the transfer of genes from *Lequerella* to rape in order to give rape the capacity for C_{20} fatty acid accumulation.

Lipid processing

Many of the process technologies used by the edible oil and chemical industries could be made selective or energy-efficient by the application of new processing techniques. Examples are indicated below:

- applications of enzyme technology in glyceride splitting for fatty acid production, the transesterification of triglycerides, and the stereoselective hydrolysis and synthesis of mono- and di-glyceride;
- the use of homogeneous catalysis in the selective hydrogenation of unsaturated oils and fatty acids;
- applications of metathesis in catalysis; co-metathesis of unsaturated fatty acid esters;
- development of novel esterification catalysts, for fatty acid esterification and for transesterification of fatty acid esters;
- development of new separation processes for the isolation and purification of sterols.

Derivatives

Chemical research into new synthetic methods is required to achieve the following reaction targets:

- shortening of C_{16}–C_{22} fatty acids to the less abundant C_8–C_{14} fatty acids;
- synthesis of specific unsaturated fatty acids by dehydrogenation of saturated fatty acids;

- introducing other functional groups (e.g. epoxy, hydroxy, carboxyl groups) into fatty acids;
- oxidative splitting of double bonds in unsaturated fatty acid esters without the use of ozone;
- synthesis of phospholipids lipoproteins with specific properties.

Biotechnology

There are many opportunities for the application of biotechnology to the production of raw materials and the manufacture of oleochemicals. These include the following:

- genetic engineering of oilseed crops to improve oil yields, oil composition and climatic range;
- use of fermentation techniques on renewable substrates by microorganisms, plant cells and animal cells, to produce unusual or entirely novel oils;
- the isolation of enzymes capable of highly selective catalysis, and their applications in glyceride hydrolysis, glyceride transesterification and glyceride synthesis; and introduction of functional groups into fatty acids.

Environmental Aspects

In this area research is needed to tackle the problems of pollution produced by processing oils and fats, notably:

- odour nuisance prevention;
- removal of heavy metals from effluents;
- treatment and destruction of residues.

The overall research situation is extremely complex. It is nevertheless evident that it is important to revitalize research into oleochemistry at both university and industry level.

FIBRE CROPS

Opportunities for fibre crops are smaller than they have been historically on account of the wide range of artificial fibres that can now be produced by the chemical industry from fossil resources (Neenan, 1983). Plant fibres offer significant advantages over artificial fibres, notably their strength and ability to stretch, as Table 2.6 shows.

It is principally for environmental reasons that interest in fibre crop production is on the increase. The main crops grown worldwide

Table 2.6. Plant and artificial fibres compared.

Fibre	(GPa) Tensile strength	Specific strength = strength value/specific gravity GN m^{-2}	Cost ratio (wood = 1)
Wood	1.2	0.8–2.7	1
Straw	0.45	0.3–1.0	0.6
Kevlar	2.87	2.0	18
Elephant glass	3.50	1.3	6

Source: Bolton (1992)

Table 2.7. Production of the main fibre crops, 1990/1000 Mt.

Crop	World	Europe	North America	Asia
Cotton	18,457	331	3653	8826
Flax	830	215	–	237
Hemp	214	49	–	136
Sisal	378	–	67	16
Jute	3626	–	11	3512

Source: FAO Yearbook (1991)

for fibre production are cotton, hemp and flax. Only cotton is a major competitor for synthetic fibre, as Table 2.7 shows.

The very small quantities of fibre crops produced in Europe only represent 2% of the arable land used. This is not just a result of the lack of interest in fibre crops, however, but also reflects the difficulty of growing some of these crops in northern climates.

Cotton is a crop of major industrial importance. It is widespread throughout the world, being grown in 90 countries. It needs a climate with hot summers; in western Europe, only Greece and southern Italy have summers warm enough for cotton production. Of the two only Greece has a significant annual production, reaching some 150,000 tonnes, and producing between 20,000 and 30,000 tonnes of cottonseed oil and 135,000 tonnes of cottonseed cake as by-products. In fact, cotton is weak and comparatively expensive compared with other plant fibre sources such as jute which is a stronger competitor. Only the breeding of cotton varieties better suited to European climates would enable its widespread production.

Flax can be grown for two completely different purposes: flax is the raw material for linen textiles, but it can also be grown for flax oil, usually called linseed oil. Of the two, the market for linen is much the larger, but it has suffered competition from synthetic fibres. In Europe flax production is small, but quite widespread; France is the main producer but flax is also grown in Belgium, Ireland, Italy and The Netherlands. There may be a new opportunity developing for the fibres from oil flax, as they constitute a possible substitute for asbestos, now banned as building material in many countries.

Hemp is grown successfully in northern climates. The annual production of hemp is now small in Europe, only about 120,000 tonnes per year. The crop was used extensively in the past for products like harvesting and baling twine and rope. These traditional markets have now been taken over by synthetic fibres such as polypropylene. For some years there was a particular market in France for hemp as a cigarette paper but this market is now also declining. One obvious problem with its production is the narcotic properties of the plant which means that its production and end-use have been banned in some countries, such as Germany.

Another important potential outlet for the fibre crops is in pulpmaking for the paper and board industry. In Sweden there is a major new government-supported project known as 'Projekt Agrofibre' (Berggren and Monties, 1991). It was started in 1987 and reflects the concern of the Swedish government about what to do with 500,000–1,000,000 ha of land which have to be taken out of food production. There was general public opposition to a massive extension of Sweden's forests, so the government sought an alternative use for the land. The project is jointly financed by the government, the farmers, and the Swedish energy and paper pulp companies. To date a total of 15 million Swedish kroner have been spent on the project. The goals of the project are to decide the following questions:

- Is it possible to produce good-quality pulp from agricultural crops?
- Is it possible to process these materials in an environmentally friendly manner?
- Is it possible to produce pulp at prices which are competitive with those for wood-based pulps?

The project thus has three components:

- pulpmaking, quality of fibres;
- methods of pretreating the raw material;
- agricultural production of the raw material.

Some of the crops tested have given high quality pulp. Examples of such crops are *Miscanthus sinensis* (elephant glass) and *Phalaris arundinacea* (reed Canary grass). Technical solutions are available for processing these crops, and the cost of the raw material is at the same level as the cost of wood with the prospect that it may be possible to reduce

Table 2.8. Productivity of fibre crops.

	Dry matter/t ha^{-1}	Unbleached pulp/t ha^{-1}
Elephant glass	15	7.8
Reed Canary grass	12	4.7
Alfalfa	10	2.6
Birch and spruce	4–5	1.7–1.8

Source: Berggren and Monties (1991)

Table 2.9. Fibre characteristics.

	Fibre length/mm	Coarseness/mg m^{-1}	Water retention value[1]/g 100g^{-1}
Pine	1.79	0.172	133
Birch	0.86	0.100	130
Eucalyptus	0.78	0.075	
Straw	0.64	0.087	180–200
Elephant glass	0.96	0.092	168
Alfalfa	0.60	0.075	210
Reed Canary grass	0.67	0.082	200

Source: Berggren and Monties (1991)
[1] Annual fibres are thin and tend to hold water more than do wood fibres. The dewatering capacity is a problem and limits the amount of annual fibres that can be used.

this. The cost of processing the crops is higher than that for wood, however, particularly where environmentally preferable processes have to be used. Trials have shown the levels of pulp production for several crops have given the results shown in Table 2.8.

Table 2.8 includes two conventional forestry sources of pulp for comparison and it can be seen that yields of the agricultural crops are significantly higher. It is not just yield and cost that matters, however, but also the quality of the fibre. For good-quality pulp, thick-walled cells are needed, and this feature is reflected in the coarseness of the fibre. Strength is an important feature of wood fibres and enables a smooth quality of paper to be achieved. Strength is reflected by both the length of the fibre and its thickness: the longer the fibre, the stronger it is. So manufacturers of, for example, cardboard boxes would look for pulp made from longer fibres. Table 2.9 gives, for comparison, the characteristics of the different fibres.

Unfortunately, in practice only the wood fibres will survive the mechanical pulpmaking process. The other processing methods available raise the cost of the pulp so far that the alternative crops to wood are not viable at present. The capital cost is also prohibitively high.

Wood

The most common basic agricultural raw material for non-food purposes is wood. It is used for a wide range of technical and chemical uses. Its most important role is as a provider of cellulose for the paper industry. Research is highly developed for this outlet but neglected as far as the other applications of wood are concerned.

Worldwide, about three billion cubic metres of wood are produced each year. Industrial countries meet about 90% of their demand from their own forests. The European Union is a substantial wood importer but most of the demand is met by other west European countries. Tree plantations can be grouped into two categories: those which are planted on a long rotation of 80–120 years, and those planted on a short rotation of 3–5 years. The application of improved breeding methods to selection, crossing and cloning have helped to raise yields. The cultivation of fast-growing varieties of poplar and willow have raised yields of dry substance to 15 t ha^{-1} year^{-1} on some experimental sites. That is five times more than the average quantity recovered from the wild.

Research objectives for wood reflect both the need to improve the raw material and the processes used to recover and make use of it. Some of the goals of genetic engineering of woods include the breeding of woods with longer cellulose fibres and trees with bark which is easily removed by mechanical means. There is an obvious need to improve on the environmental consequences of wood processing. This is well illustrated by the problems of Finland, where many of the country's lakes are chronically polluted by the wood-processing industry.

Wood can be classified as an agricultural raw material where it is grown for a purpose. Its most obvious application is as solid wood for the construction industry and furniture manufacture. A range of wood products, such as chipboard, are also used by the construction industry. The paper industry is the main user of wood pulp as well as waste wood from forestry. Wood could be used as a major chemical feedstock for the chemical industry, for it can be fermented to methanol which is a basic chemical building block.

Producing wood for energy purposes on arable land no longer needed for food production is an interesting alternative in the current agropolitical climate in Europe (Centre for Agricultural Strategy, 1992).

A tree crop which is grown for energy production needs to be as efficient as possible at binding carbon and thereby producing the maximum amount of biomass per unit of area. Selected clones of *Salix* (willows) have good growth capability and have demonstrated their capacity to provide energy corresponding to 5.5 m^3 of oil ha^{-1} year^{-1} in field trials in Sweden. For maximum returns a coppicing method is used. The willows are propagated by means of cuttings which are 25 cm long unrooted sticks. They are planted directly into weed-free ground in the spring at close spacing (10,000 trees ha^{-1}) by either manual or mechanical means. Traditional transplanting equipment has been successfully adapted for this purpose. The establishment rates are typically high (in the region of 90–95%) where weed-free conditions are maintained and sufficient rainfall occurs after planting. At the end of the first growing season, the shoots produced are cut back to initiate coppicing. The cut shoots can themselves be used as cuttings for subsequent planting.

In the first season it is important to keep the field conditions weed-free to limit the competition for the available water. The first harvest can be undertaken 2–5 years from the initiation of the coppice. The regular harvest can then be made every 3–5 years up to an optimum maximum age for the coppice of 30 years. By planting a field in sections and rotating the crop between them it is possible for the farmer to achieve an annual harvest and therefore income.

Harvesting is carried out in the winter, or any time after leaf fall, by mechanical means. Harvesting equipment is still being adapted for this purpose. The Department of Energy in the United Kingdom has developed a prototype which can be tractor-drawn. It uses contra-rotating augers to gather the stems, which are cut close to the base using a circular saw. The cut stems are collected in a baling chamber and the bales are emitted from the machine from time to time. These bales of cut stems can be easily stored for subsequent chipping and burning round the year.

What is left of the tree after harvesting is known as a cut 'stool'. These appear to be remarkably resilient to mud, frost and tractor tyre damage and usually re-sprout in the spring. During its 30-year optimum production period the size of the stool will reach a plateau and decline towards the end of its life. So productive are the roots of an active stool that weed control is not usually a problem after the first two years of establishment.

It is difficult to ascertain the cost of arable energy forestry accurately by this method. It depends on several variable items: fencing, ground preparation, cuttings, harvesting, transportation and chipping. The UK Department of Energy has tried to establish the cost for a 10 ha site (Maryan, 1992). Ten thousand trees were planted per hectare on three equally sized plots in successive years, giving a harvest every three years. The field was permanently fenced against rabbits. Cuttings were purchased from a UK supplier at 20p each (in Sweden, cuttings are available more cheaply – at about 8p each). Cuttings have to be bought only in the first year, as the cutback from the first plantings can be used subsequently. These were valued at 5p each. The yield of the crop was estimated at 12 dry t ha^{-1} year^{-1} (the average result from field trials to date). Harvesting and chipping were carried out under contract.

In the UK arable energy forestry is eligible for the Forestry Commission Woodland Grant Scheme and the 'Better Land Supplement'. Taking all these factors into account the cost was estimated as follows:

	£
Fencing at £3 per metre	3800
Cuttings at 20p for 3.3 ha and 5p for 66 ha	9900
Site preparation	600
Weed control	650
Total	£14,950

These costs reflect the set-up cost of the coppice over a three-year period before income is possible from harvesting. However, there is grant income available as follows:

	£
10 ha under Woodland Grant Scheme	9750
10 ha Better Land Supplement	6000
Total	£15,750

Although this scenario would provide a small surplus of £800 at the end of three years it should be noted that the Woodland Grant Scheme is phased in over a three-year period with 70% paid in year one, 20% in year five and 10% in year ten. This would mean that there might be periods of deficit when interest charges would be incurred. Nevertheless it is broadly safe to say that the cost of establishing the crop can be met by grant income.

All the other costs of producing the fuel have to be calculated at contractor rates. The cost of contractor harvesting and chipping of coppice wood based on field trial data is £19 per dry tonne, but this can be reduced to £9 per tonne if harvester and chipper are operated to their designed productivity. The cost of the fuel wood and the cost per gigajoule (GJ) are shown in Table 2.10.

The 'income' quoted in Table 2.10 is not a gross margin or an annual

Table 2.10. Costs of fuel wood from coppiced woodland.

'Income' (whole field)	Fuel wood/£ dry t^{-1}	£ GJ^{-1}
0	19.00	0.97
100	27.40	1.40
200	35.80	1.84
300	44.21	2.27
400	52.61	2.70

Source: Energy Technology Support Unit (1991)

Table 2.11. Comparative costs of fossil fuels.

Fuel	£ GJ^{-1}
Natural gas	2.99
Gas oil	3.93
Light oil	2.36
Medium oil	1.95
Heavy oil	1.48
Propane	7.30
Butane	5.79
Coal (singles)	2.34
Coal (smalls)	2.25
Electricity (off-peak)	6.11

Source: Maryan (1992)

equivalent value. This is because the farmer's fixed costs cannot be calculated without reference to his or her individual situation. However, it is clear that the management input to the crop is low and can be carried out at a slack period for the arable farmer. The sort of income a farmer could expect is calculated by reference to competing fuel prices. Table 2.11 shows the average gross price of fossil fuels to industrial users.

If transport and storage costs are added to the farm gate price of wood fuel and account taken of the capital cost of the wood-burning plant, its cost to users is about £2 GJ^{-1}.

Some alternatives to the coppicing method are to use the by-products of conventional forestry operations and wood product manufacturing, or to use wood in 'derelict' woodlands.

The main aim of forestry is to produce timber and timber products from the tree. In general they are derived from the stem of the tree, and forestry management practices are designed to give the maximum production of knot-free stemwood. Mechanical harvesters are increasingly being used to fell forestry plantations, but some chainsaws are still in use and also still tend to be used for pruning. Since it is only the stem of the tree which is removed, the other parts, and notably the treetops or 'brash', are left on the ground. The brash accounts for some 30% of the total tree biomass and can be perfectly well used for energy production. The simplest ways of harvesting it are to re-enter the site and either collect the wood and chip it on site using a mobile terrain chipper, or remove it to the woodside for chipping in a static chipper. Unfortunately, these residue harvesting methods are not very efficient and produce energy at a cost of £2.18–2.33 GJ^{-1}. An integrated harvesting operation consisting of one pass could reduce this cost of producing one gigajoule of energy to £1.17 GJ^{-1}. The drawback with integrated harvesting systems is the high capital costs associated with the specialized equipment.

Another potential source of fuel wood from forestry operations is that recovered after essential thinning. In this case, whole tree chipping makes economic sense. With this system productivity is high and the price for delivered fuel wood is in the region of £1.79 GJ^{-1}.

Wood fuel is available now at a price which is becoming increasingly competitive with fossil fuels. Arable energy forestry can produce fuel wood and remove land from food production. The coppicing methods tested are highly productive and the economics of the crop are improved by the forestry grants available to get the woodland established. The crop can remain in production for 30 years with low inputs in terms of both time and agrochemicals. Fuel wood can be supplied on long-term contracts at stable prices, which offers a further significant advantage over fossil fuels.

SECONDARY SUBSTANCES

The range of plant products known as secondary metabolites includes fragrances, colourings, intoxicants, stimulants and even poisonous substances. Growth in this area is being fuelled by the consumer movement towards natural products and away from artificial additives. Well-known examples of these plants include camomile, peppermint, valerian, lavender, pine, poppy and foxglove. A major market opportunity exists for dedicated producers supplying users on a long term basis (Table 2.12).

Essential oils such as peppermint and camomile used to be produced in Europe but were phased out on account of the production cost. Also labour-saving specific herbicides were not cost-effective to produce. However, research into labour-saving techniques and trials

Table 2.12. Worldwide merchant sales of fragrances and flavours 1984–90.

Product	1984		1990		Growth Rate 1984–90/%
	Sales/ US$ m	Share/ %	Sales/ US$ m	Share/ %	
Fragrance compounds	1650	35.5	2400	35	6.5
Flavour compounds	1500	32.3	2300	33.6	7.5
Aroma chemicals	750	16.1	1100	16.1	6.5
Essential oils	750	16.1	1050	15.3	5.5
World total	4650		6850		6.5

Source: Caiger (1993)

with conventional herbicides have given fresh impetus to the production of these essential oils in Europe. For example, it has been found that peppermint can be harvested with a conventional forage crop harvester. This shows how modern technology can make it possible to re-introduce crops ruled out previously on account of their high labour costs. There is an enormous variation in the value of these unusual crops ranging, for example, from US$800–1000 kg^{-1} for camomile oil to US$10 kg^{-1} for eucalyptus oil. This is now a growing market, estimated to be worth US$200 million worldwide.

A comparatively new area consists of the commercial production of wild flowers. These are used by the landscaping industry and are particularly popular where a project has a bias towards environmental protection. The initial cost of transplanting wild flowers was prohibitive, being of the order of 50p per plant compared to 8p for a conventional vegetable transplant. However, technological improvements have brought the cost of transplanting wild flowers down to 12–20p per plant. Many of these wild flowers have found a value-added outlet as they are dried and preserved.

Another new opportunity has arisen for agricultural raw materials in the pharmaceutical sector. Many plants have reputed therapeutic properties and a few of these can be turned into commercial opportunities. Attitudes towards the use of plant remedies have turned full circle over the last 30 years. In the 1960s opposition grew to plant-based pharmaceuticals and attention turned instead to chemically derived drugs. However, there were a number of bad experiences with the side-effects of synthetic chemicals, such as the Thalidomide disaster. The chemical approach to medicine has failed to deliver cures for many chronic diseases. The increase in the use of biotechnology in medicine has placed the emphasis back on 'natural' approaches to therapy (homeopathy). Finally the increasing concern of the general public about the environment and pollution has increased scepticism towards chemical treatments generally.

There are numerous examples of plants with therapeutic properties. One of the most recent successes has been the use of evening primrose to treat eczema. About 1% of the population suffer from atopic eczema, a painful skin condition resulting from their inability to convert linoleic acid from fats such as margarine. The seed oil of evening primrose contains γ-linoleic acid and, if taken in a daily dose of 250–1500 mg, it can significantly improve the symptoms of eczema. People with diabetes also have difficulty converting linoleic acid from food and this can ultimately cause nerve damage. Thus the 'market' for this product may be even wider. Evening primrose fulfils the criteria by which a plant product can be judged to be commercially viable:

- Is there a genuine therapeutic need?
- Is the therapeutic activity due to a single substance since governments are reluctant to grant licenses for multiple activity ingredients?
- Is it reasonably easy to isolate the active material from the plant source in a consistent form?
- Is the plant feasible from an agronomic point of view?
- Is there any possibility of intellectual property protection by patents or other means?
- Can it be produced at a reasonable cost?
- Is it safe?

In the case of evening primrose it would be feasible to obtain a patent under the US system, and the patent system of the European Community has now broadly come into line with that of the USA – in fact, it probably offers even stronger protection. The pharmaceutical industry estimates that it costs £150m to develop a new drug, of which £15–25m is spent on medical research, £20m for government drug licence and another £15–25m on sales and marketing. By comparison crop developments costs are small, at only £1–2m.

ENERGY FROM WASTE PRODUCTS

Another somewhat unlikely raw material for non-food use consists of agricultural or urban waste. The central feature of waste treatment is that it deals with very large quantities of material at extremely low unit cost. The most efficient methane-producing system in nature is the animal rumen, where a complex system of anaerobic digestion takes place. The principle from this natural process has been replicated quite effectively for commercial purposes. The anaerobic digestion of sewage is a technique long practised by municipal authorities for capturing methane and harnessing it for the energy needs of the sewage-treatment system.

Plenty of unutilized urban and agricultural waste is available, but its use for energy production has been hindered by economic considerations. A particular problem of sewage waste treatment is the disposal of the residual sludge, which can represent 40% of the total treatment cost (Dunnil and Rudd, 1984). Around 30–50% of the convertible energy can be recovered from urban waste, and with agricultural waste this figure can reach about 70% (Smith, 1988). Vegetable waste material and forages make the best agricultural substrates. The by-products generated by the process are rich in ammonia, phosphates and microbial cells and as such make good fertilizer, soil conditioner or even animal feed. The economic problems are related to the cost of collecting the organic matter and the irregular rate of methane production. Another practical problem in using agricultural waste is that lignins are not converted by anaerobic digestion; indeed, much research is still needed to optimize this process. Some experiments have been carried out using crops specifically grown for the purpose of obtaining a high yield of methane, such as water hyacinth, but they have not expanded to large-scale commercial use.

The recovery of methane from animal dung is referred to as *biogas* production. It is a near-anaerobic process rather than one requiring the total exclusion of oxygen. It is a very old practice and of particular importance in the Far East. Most biogas generation plants are small, and thousands are in operation throughout the world at family, farm or village size.

Conclusion

This chapter may not have provided an exhaustive list of all the agricultural raw materials and plant varieties which can be used for non-food purposes, but it has attempted to identify the main opportunities for farmers to diversify into non-food production. The leading commercial opportunities are for carbohydrates, oils and fats, and fibres from well-known existing commercial crops. But the opportunities for farmers to diversify their crop base to incorporate less conventional crops such as those used to produce secondary substances should not be overlooked. An individual farmer who manages to secure a contract to produce a specialized crop such as evening primrose for pharmaceutical purposes may find this making more than a marginal contribution to on-farm profitability.

Straw and wood are among the less conventional agricultural raw materials still handicapped by their lack of economic viability compared to their competitors. As has been discussed, the obstacles to their wider use are in part technical and it may be hoped that new techniques are found to overcome these hurdles. The new technologies such as biotechnology offer the prospect of technical breakthrough by enabling completely new processing techniques to be employed or the characteristics of the plant itself to be altered. The more widespread use of less conventional crops for non-food use would be greatly assisted by government schemes which favour their introduction and support research and development. The European Community has taken several initiatives on this front, which should help to encourage pre-competitive research for non-food crops. Even more may have to be done if the fruits of this research are to be seen in the market entry of these crops.

There are broadly three domains of opportunity for agricultural raw materials for industrial use:

- the replacement of imported oils, e.g. castor oil;
- the modification of products to extend their markets;
- the creation of new products, e.g. biodegradable plastic.

One of the main factors driving substitution, modification or the search for completely new products is concern for the environmental damage caused by conventional products and processes. Many agricultural raw materials offer distinct advantages over their mineral equivalents and thus it must be anticipated that use of the former will expand at the expense of the latter.

REFERENCES

Berggren, H. and Monties, B. (1991) *Projekt Agro-fibre*. Paper presented at the International Centre for Under-utilised Crops Symposium, 12–13.12.91. School of Pharmacy, University of London.

Bolton, A.J. (1992) *Fibre Bonding and New Uses of Plant Fibre*. The Biocomposites Centre, University of Wales, Bangor. Paper given at LINK Seminar on crops for industrial use, London, 18.5.92.

Caiger, S. (1993) Markets and opportunities for alternative high value horticultural crops: strategies for development. In: Anthony, K.R.M., Meadley, J. and Roebbelen, G. (eds), *New Crops for Temperate Regions*. Chapman and Hall, London.

Centre for Agricultural Strategy (1992) *Energy Coppice: an Alternative Farm Crop*, Issue 2. University of Reading, Reading.

Council of European Chemical Manufacturers Federations (CEFIC), (1985) *The*

Use of Agricultural Raw Materials in the European Chemical Industry. Brussels.

Dordick, J. (1991) *Biodegradable Sugar-based Polymer.* World International Property Organisation Publication Number WO/17255.

Dunnil, P. and Rudd, M. (1984) *Biotechnology and British Industry.* A report to Science and Engineering Research Council of the UK (SERC), Polaris House, North Star Avenue, Swindon.

European Federation of Oilseed Crushers and Processors (1991) *Table of Marine and Vegetable Oils and Fats.* General Assembly, Brussels.

Energy Technology Support Unit (ETSU) (1991) *Arable Coppice.* Harwell Laboratory, Oxon.

Food and Agriculture Organization (FAO) (1991) *FAO Yearbook.* FAO, Rome.

Kollonitsch, V. (1970) *Sucrose Chemicals.* International Sugar Research Foundation.

Lichtenhalter, F. (ed.) (1990) *Carbohydrates as Organic Raw Materials.* VCH, Weinheim.

Pennington, N. and Baker, C. (1990) *Sugar, a User's Guide to Sucrose.* Van Nostrand Reinhold, New York.

Maryan, P.S. (1992) *Wood Fuel Production.* ETSU, Harwell Laboratory, Oxon.

Neenan, N. (1983) *Textile Fibres and Related Products.* The Agricultural Institute, Carlow, Ireland.

Shields, R. (1992) *Industrial Oils from Crops.* Plant Breeding International Ltd, Cambridge. Paper given at LINK Seminar on crops for industrial use, London, 18.5.92.

Smith, J.E. (1988) *Biotechnology,* 2nd edn. Edward Arnold, London.

3

TECHNOLOGY

The purpose of this chapter is to examine the technological aspects of using agricultural raw materials for non-food purposes. A wide variety of technologies is employed in non-food production; these range from conventional manual techniques used for many centuries through to the latest technology resulting from the application of the most recent research into the molecular structure of living organisms.

The conventional methods of processing agricultural raw materials for non-food purposes are physical processes, using either force or heat to break down the raw material and obtain the required characteristics of the end-product. Many manual techniques are still used to cut, roll, crush and grind agricultural raw materials in order to extract the required product. They can often be seen in use in developing countries, where agricultural fibres in particular are still heavily used for construction purposes and for everyday utensils. In the developed world many of these processes have been mechanized to reduce the labour costs but the machines fundamentally undertake the same tasks of cutting, grinding and so forth. One of the major limiting factors on the expansion of non-food uses in developing countries is the cost of the new technology.

There are several reasons why new techniques developed in countries which are advanced economically are so different from those used in less-developed countries (LDCs). For some years there has been an extensive debate about the type of technology which is desirable for the under-developed world. A machine which reduces labour costs but which is complex to maintain and costly to repair may not be the appropriate technology at all. This fact has led to the introduction of the term *intermediate technology* to describe technology that is neither so complex and expensive nor so primitive that the available resources in the developing world cannot be used optimally. The wheel may yet turn full circle as developed countries are prevented from using some of the advanced techniques they have perfected on health or environmental grounds. Some of the more 'old-fashioned' techniques may prove more environmentally benign.

More advanced techniques involve the use of plant-processing equipment and heat to extract the agricultural raw material required. These range from relatively low-temperature techniques used to melt or liquefy solid ingredients through to high-temperature methods such as distillation and crystallization used to break down the cellular structure of the organism to extract the required product and then refine it to the required degree of purity. Sometimes chemicals are employed on their own or in conjunction with heat to achieve the same end. Chemicals can speed up a natural process; for example, in the case of the retting of flax fibres the natural elements would make the flax fibres available for use with time, but the application of chemicals to a flax field will speed up this process (Marshall, 1989). Similarly, the addition of a cocktail of chemicals to the extracts of sugar beet or cane will help separate the sugar from the other impurities, leaving less of it behind in the molasses.

New technical possibilities for non-food uses have been made possible by the advent of biotechnology. In the rest of this chapter we will concentrate on this new enabling technology and the opportunities it creates.

BIOTECHNOLOGY DEFINED

The development of this particular technology over the last ten years has made the wider use of agricultural raw materials possible. Given the novelty of these developments, it is not surprising that there is considerable variation in the way in which biotechnology is defined. The Office of Technology Assessment of the United States Congress (1989) undertook a study of all the available definitions. As a result, it was possible to synthesize the various definitions into the following:

> Biotechnology, broadly defined, includes any technique that uses living organisms (or substances from these organisms) to make or modify a product, to improve plants or animals, or to develop microorganisms for specific uses.
>
> (OTA, 1989)

The OTA attempts a refinement by making a distinction between 'old' and 'new' biotechnology. By 'old technology' is understood the biological processes known to humans for many centuries such as brewing and bread- and cheese-making. By 'new technology' is understood revolutionary new techniques such as genetic engineering. The Dutch government, for example, is quite categoric about what types of technology cannot be called biotechnology; in their view 'medical technology, agriculture, and traditional crop breeding are not generally regarded as biotechnology'. Their definition concentrates on 'the science of the production processes based on the action of microorganisms and their active components'. However, such a definition has its limitations.

The OTA definition has the virtue of being broad enough to embrace both the precommercial and the market applications of biotechnology. Several of the other definitions of biotechnology stress one or other of these aspects. The Organization for Economic Co-operation and Development (OECD) puts the emphasis in its definition upon the application of biotechnology:

> Biotechnology consists of the application of scientific and engineering principles to the processing of materials by biological agents to provide goods and services.
>
> (OECD, 1989)

A good definition which places biotechnology in the context of scientific progress is given in Persley (1990):

> Biotechnology is (thus) comprised of a continuum of technologies, ranging from the long-established, and widely used technologies, which are based on the commercial use of microbes and other living organisms, through to the more strategic research on genetic engineering of plants and animals.

This definition emphasizes that biotechnology is not an entirely new science but rather an extension of existing technology that has opened the way to completely new techniques in handling living organisms.

The Technologies

As can be discerned from the problem of definition, there is in fact more than one technology. For this reason, we may speak more correctly of the biotechnologies. Under this umbrella, we can find several well-known new techniques for making and modifying organisms. Amongst the most famous is *recombinant DNA technology* (rDNA). Deoxyribonucleic acid (DNA) is the carrier of genetic information in chromosomes found in all living organisms. rDNA techniques permit the formation of a new association of genetic information which will be expressed as different characteristics from those of the parent organisms. It allows the direct manipulation of the genetic material of individual cells. Over the last ten years, several important discoveries were made in molecular biology which led to the development of rDNA technology. Of particular significance was the identification of a strain of the bacterium *Escherichia coli*, the restriction enzymes of which are able to break down foreign DNA entering a cell by cutting the DNA at specific sites or 'restricting' it. The use of this and other restriction enzymes allows the manipulation of DNA and the creation of new arrangements of genetic material.

PLANT BIOTECHNOLOGY

Over the years, plant breeders have been using conventional methods to increase crop yields and to increase resistance to disease, pests and environmental conditions, as well as to improve the quality of crops. Biotechnology offers a powerful new tool to accelerate and render more accurate the process of breeding. Some of the practical and commercial benefits of applying these new technologies may eventually include disease resistance, enhanced photosynthetic efficiency and improved nitrogen fixation. These developments will clearly help to reduce the costs of production in agriculture and are therefore worthy of more detailed consideration.

Plant Genetic Engineering

DNA is a double-stranded helical molecule, the backbone of which is composed of four nucleotide bases: adenine (A), cytosine (C), guanine (G) and thymine (T). A gene is a part of a DNA molecule which contains an ordered sequence of these letters along one strand with a complementary sequence on the second strand. Normally, DNA will replicate itself perfectly by separation of the strands, each of which then produces a further mirror image sequence. However, it is possible to cut the original sequence of DNA and insert a new section of DNA including a new gene of interest. Once this DNA is recombined and replicated, the new organism will give expression to the new gene within it. By this means it is possible, for example, to obtain new characteristics in commercial crops. Another very important new technique is *cell fusion*, by which cells are joined together and the desirable characteristics of different cells are combined. In theory, the fusion of cells from two different plant species can be used to overcome barriers between interspecies hybridization; protoplasts (cells without their cell walls) are used to facilitate fusion. In order to be useful commercially the cells resulting from this technique must be able to regenerate and form a whole plant; however, at present it has only been possible to regenerate plants resulting from the fusion of related parent material. Cell fusion of animal species has been more successful commercially as, for example, in the production of monoclonal antibodies from the new cell line following a fusion which can be used in the diagnosis and treatment of disease. Monoclonal antibody formation is carried out by injecting a mouse or rabbit with the antigen, removing the spleen and then allowing the fusion of spleen cells with myeloma cells. 10% of the final hybridomas are antibody secreting cells which secrete one antibody.

Another important area in which biotechnology has contributed significant commercial advantages to agriculture has been in *bioprocess engineering*. Whilst the techniques used are not novel genetic ones, they do allow the adaptation of biological methods of production to large-scale industrial use. One example of this is plant cell culture, by which whole plants can be regenerated from a selection of virus-free cells. Yields of commercial crops are substantially reduced by plant viruses and in the past plant breeders only had trial and error methods of selecting for virus-free material. Plant cell culture acts as a bridge between molecular genetics and plant breeding. In the research laboratory single plant cells can be examined, selected and cultured to produce whole plants. Plant cells have the characteristic of totipotency, that is, each one carries all the genetic information needed to make a complete plant. If the same number of whole plants were used in an experiment, not only would the process take much longer but many hectares of land would be required. As mentioned earlier, the regeneration of plant cells still poses some problems to researchers. Regeneration is achieved by growing an undifferentiated clump of cells, known as a callus, from a single selected cell under controlled conditions of light, temperature and nutrients, including growth-regulating chemicals. Then either plant embryos can be formed from the callus, or the callus is allowed to differentiate into plant root and stem tissue. Regeneration has been successful in dicotyledons such as rapeseed, sunflower, potatoes, tomatoes and sugarbeet, and has recently been achieved in some cereal crops.

An important opportunity is afforded by the callus-growing technique to insert new genetic material into the plant cells (Fraley, 1988). In the past, the introduction of new genetic information into plant cells was achieved by the induction of mutations (changes in genes or chromosomes) using physical or chemical means. However, this method is somewhat haphazard with limited control over the resulting capabilities of the plant. The variation in the characteristics resulting from mutation is known as somoclonal variation. Such mutations can lead to small improvements in product formation (5–10%). *In vitro* selection with a herbicide or pathogen toxin can be performed to identify those cells preferentially resistant to the agent, allowing regeneration of a resistant plant. A more accurate method of introducing genetic information is by the insertion of specific genetic information into the plant. Scientists discovered that the bacterium *Agrobacterium tumefaciens* contained a plasmid (a minute fragment of nucleic acid) capable of inducing tumours. This tumour-inducing (Ti) plasmid is a natural genetic engineer as it inserts genetic information into the DNA of a plant and the desired gene can be inserted into the plasmid (Grant *et al.*, 1991).

A similar technique also uses the *Agrobacterium* genus, this time with a root-inducing plasmid (Ri). (It is necessary, of course, to remove the disease-causing aspects of the Ti and Ri plasmids in order that the disease does not appear in the resultant plants or any plant material regenerated from them.) There have been many successful applications of this technique. One of the first examples of this was the transfer of a bacterial gene for antibiotic resistance into pieces of a plant leaf. The engineered leaf pieces were then cultivated on a medium containing the antibiotic that would normally kill them, but they survived. This constituted a breakthrough in research, showing for the first time that a bacterial gene will function in a plant cell. DNA can also now be introduced via tiny holes in the plant cell wall created using electric charges (electroporation, see Rathus and Birch, 1991), by chemical means or by firing metal particles coated with DNA into the cells (Franks and Birch, 1991).

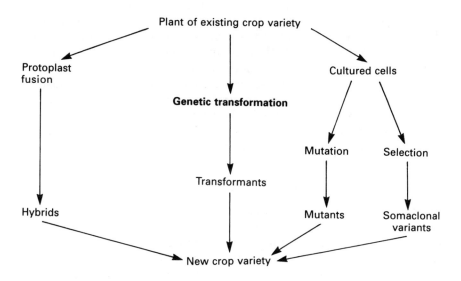

Fig. 3.1 Alternative methods of obtaining new crop varieties (OTA, 1984).

Resistance to Disease

Lack of resistance to disease caused by fungi, bacteria, and viruses results in substantial losses of productivity. Higher plants are known to contain genes for resistance. Traditionally, disease resistance has been bred into plants by interbreeding different species. However, biotechnology makes it possible to clone disease-resistance genes (the genes themselves are not resistant but they code for resistance) to study the possibility of transferring them between species that will not usually interbreed. Research is still at a preliminary stage in understanding what disease-resistance genes do to a plant's metabolism and structure. One of the reasons for the lack of progress in this area has been the attraction to the large agrochemical companies involved of exploiting the commercial benefits of an alternative route – for example, to develop plants that are pesticide-resistant rather than pest-resistant. It may be argued that it is more desirable to develop pest-resistant plants because this would reduce the need for spraying crops with chemicals, but this offers less incentive for these companies to pursue this route.

This observation must be tempered by the fact that research efforts have yielded up progress in viral resistance among plants. Genetic engineering has made it possible to 'vaccinate' plants against certain viruses. An example of this is the work done on cucumber mosaic virus, which affects a diverse group of plants including vegetables such as cucumbers, tomatoes, peppers and other commercial crops. When plants susceptible to the virus are inoculated with the coat protein of the virus, the development of the symptom and accumulation of the virus is reduced or absent in the subsequent generations of offspring of the inoculated parent plants. Mechanical inoculation of plants can, however, prove difficult. Therefore researchers have resorted to the use of *Agrobacterium tumefaciens* bearing the gene for coat protein to inoculate plants more accurately. This procedure of *agroinfection* has also proved more practical than the attempts to eradicate the insect vectors that transmit many mosaic virus diseases in the wild.

Resistance to Herbicides

Genetic engineering can help plants to develop resistance to chemicals. Some plants do already have natural resistance to certain herbicides; for example, maize is naturally resistant to triazine herbicides. Resistance traits take several forms: either the plant does not absorb the herbicide, or it does not transport the herbicide within itself, or it may detoxify the herbicide. A traditional method of breeding in resistance is to cross commercial crops with resistant weeds, but this tends to result in lowered yields.

The research and development necessary to generate herbicide-tolerant plants has in most cases evolved from studies to understand how herbicides affect the plant's metabolism. Several companies have engineered plants to be tolerant to glyphosate (Benbrook, 1986), a non-specific herbicide which kills most green, actively growing plants.

As herbicides go, glyphosate is viewed as relatively benign for the environment, as it breaks down quickly in the soil. It acts on the plant by inhibiting the production of an enzyme called EPSP synthase, which the plant needs for the production of amino acids essential for it to thrive. Using the *Agrobacterium* system, scientists inserted genes for producing glyphosate-resistant EPSP synthase, derived from bacteria, into plants so that they would produce a functional version of the essential enzyme. The offspring of these transformed plants were found to be two or three times more tolerant to glyphosate than untreated plants. As a result of the breakthrough Calgene Inc. expects to commercialize glyphosate-tolerant crops in the 1990s. Research is continuing into engineering resistance to other common herbicides such as sulfonylureas, imidazolinones and phosphinothricin (see Mullineaux, 1992, for further information).

This breakthrough with a broad-spectrum herbicide may save the farmer having to use a host of more specific chemicals. But this direction of research is not without dangers, for apart from the normal environment hazards of toxicity, soil leaching and ground water contamination, there is a risk of conferring herbicide resistance to weed species and the introduction of undesirable agronomic characteristics associated with the tolerance trait itself. The cost and risk involved in registering herbicide-tolerant crops is likely to promote a trend towards commercializing fewer, but more effective and more environmentally acceptable products. At the rate at which biochemical technology and techniques for genetic engineering are progressing, it is expected that resistance genes will be created in a test-tube by mutation or DNA synthesis.

Resistance to Insects

Some species of plant naturally produce compounds that are toxic to certain insect species. Researchers have begun to examine these characteristics of plants in order to clone and transfer the genes and develop insect-resistant varieties. An example of this is the cowpea trypsin inhibitor gene, which has been engineered into several crops (Gatehouse *et al.*, 1992).

An alternative approach involves a protein constituent of *Bacillus thuringiensis* (*Bt*), which is a naturally occurring microbe that lives in the soil. When an insect such as a caterpillar ingests *Bt* the protein is turned into an insect-specific toxin by enzymes in the caterpillar's stomach, causing paralysis and death of the organism. Since the enzymes are unique to certain larvae and not found in the stomachs of other insects or animals, *Bt* poses no threat to other life forms. In fact, *Bt* has been a widely used ingredient of home and commercial garden sprays for over 20 years. It has now been possible to engineer a gene for *Bt* protein into tomato and tobacco plants. The insertion of the *Bt* gene into these commercial crops offers the farmer a less costly and environmentally preferable method of insect control. There are, however, concerns about insects acquiring resistance to *Bt*, and some evidence of this has been reported for the diamondback moth.

Allelopathic Treatment

Some plants release chemicals which adversely affect their neighbours. These naturally occurring herbicides have not yet been fully exploited for their commercial potential. They are known as *allelopathic* chemicals and act by inhibiting cell division, protein synthesis and photosynthesis of plants. One of the objectives of researchers working in this area is to identify the genes responsible for the release of these plant herbicides and transfer them to non-resistant plants. If they are successful, another tool for alternative weed control would be placed in the hands of the farmer.

Nitrogen Fixation

All plants need to metabolize usable nitrogen for healthy growth and plant development. A few plants, such as legumes, are capable of fixing free nitrogen in the soil on account of a symbiotic relationship with *Rhizobium*, a nitrogen-fixing bacterium that inhabits nodules on their roots. To overcome the inability of most crops to fix nitrogen, farmers have traditionally applied chemical nitrogen fertilizers at key stages in the crop's development. Concern is now growing about the damage caused to the environment by this practice, in particular the effects of nitrates leaching into the water supply. In response to public concern, the European Commission has introduced a new EC water quality directive limiting the maximum allowable concentration for nitrates in drinking water to 50 mg l^{-1}. It has been demonstrated that above this level there are risks to human health of cancer and, in young babies, the risk of impairing the oxygen-carrying capacity of their blood (methaemoglobinaemia) (Owen, 1986). For these reasons, attention is now being focused on alternative biotechnological methods for fixing nitrogen.

There are two principal possibilities. One is to extend the host range of *Rhizobium* and that of another nitrogen-fixing microorganism,

Frankia. However, little is understood about nodule formation and nitrogen assimilation in this symbiotic relationship. It has proved particularly difficult to extend the symbiotic relationship to unrelated plant genera. Another possibility is the transfer by genetic means of the nitrogen-fixing ability of free living organisms, such as blue-green algae. The best studied nitrogen-fixing bacteria are *Klebsiella pneumoniae, Azosporillum* and *Azotobacter,* which can all be grown easily in the laboratory. One of the difficulties of conducting research in this area is the genetic complexity of the code for nitrogen-fixing ability, which consists of 17 different genes. The regulation and activities of these genes are now the subject of extensive research. Geneticists not only wish to transfer the genetic nitrogen-fixing ability to other plant genera but also are striving to increase the efficiency of the nitrogen-fixing process, so that plants will not suffer a reduction in yields through adaptation to fix nitrogen rather than relying upon the application of chemical fertilizers.

Seed Improvement

The technologies which constitute the new plant genetics, such as plant tissue culture and recombinant DNA technology, will radically affect the global seed business. As farmers become increasingly aware of the superior performance of seed produced by biotechnological means, demand for it will increase. There is likely to be a switch away from farm-saved seed to commercially-derived seed which assures higher varietal purity and higher mechanical quality. Most biotechnologically produced varieties will be hybrids requiring the farmer to purchase new seed each year; indeed, patents may prevent the use of farm-saved seed. These factors will lead to higher yields and thus higher returns.

Biotechnological production methods require a greater level of investment in research, production and seed conditioning than do conventional breeding methods. Therefore, it is the larger commercial companies which are better placed to invest and compete in the global seed market. This fact is likely to have a strong influence on the structure of the global seed business. In the past, there were many small seed companies, often family businesses, who rode the roller-coaster of good and bad fortune in breeding by traditional methods. One year the variety of a particular seed house may be top of the list, in great demand from all growers; a few years later it would have slumped to the bottom, out-performed by other new varieties.

Since the advent of biotechnology, there has been a startling switch to corporate investment in agricultural research, particularly by firms who produce agrochemicals. There are numerous examples of this: for example, the British family seed business, Nickersons, has been purchased by the large seed company, Limagrain. ZENECA (formerly ICI) is an example of a chemical company which has consolidated its investment in the seeds business. It is predicted that 12 large multinational companies such as Sandoz, Ciba-Geigy and Pioneer Hi-bred will dominate the global seed business shortly after the year 2000 (Kidd, 1985). The incentive for agrochemical companies to engage in the seeds market is high, in view of the fact that the introduction of plants not needing chemicals, such as nitrogen-fixing and pest-resistant plants, is likely to reduce the market for agricultural chemicals.

Another sector motivated to invest in agricultural research and plant biotechnology is the food and pharmaceutical sector, as plants can be adapted or tailor-made to meet their needs. Obviously, these developments will have an important impact on the unit costs of seed. Growers are expecting to have to pay more for seed produced by these new techniques. But it is hoped that the improved performance produced by these biotechnological means will justify the increase in seed cost. Growers may not be aware that the seed they are purchasing is a product of biotechnology; their decision to purchase or not will be a straightforward judgement between the performance of the variety in seed trials and its cost compared to other available varieties.

Plant 'Factories'

If the first area of application of plant biotechnology could be described as the capacity to improve agronomic traits such as disease resistance and the second wave as the improvement of traits affecting food processing, then a 'third wave' has now been discovered (Moffat, 1992). Early in 1989 it was found that plants could be used to produce speciality chemicals and polymers. These products were being produced by genetically engineered bacteria, but it was discovered that plants could substantially increase the yield of these products. This is specially important where the product is required in bulk – by the tonne rather than by the gram. Research has shown that plants can manufacture certain human and mouse proteins that have potential applications in therapeutic and diagnostic medicine. These discoveries have often been made virtually by accident. For example, in 1989 when Gus de Zoeten of Michigan State University and Thomas Hohn, then at the Friedrich Miescher Institute in Basel, introduced the human gene for interferon production into a turnip in order to try and reduce the turnip's susceptibility to viral infections, they discovered that their transgenic plants were making large amounts of the interferon. Almost simultaneously, workers at the Scripps Research Institute at La Jolla, California, discovered that tobacco plants could be genetically engineered to produce antibodies, or 'planti-bodies' as the researchers call them. These antibodies behave normally in laboratory tests. Some researchers have deliberately set out to see whether plants could be used as bioreactors for manufacturing human biological products. Not only tobacco and turnip plants but also potato, oilseed rape and a small plant called *Arabidopsis* have proved useful in experiments of this kind. A Dutch agricultural biotechnology company, Mogen International, genetically transformed plants to produce human albumin solutions, which are widely used for fluid replacement in burns cases.

Despite the importance of these discoveries for the field of medicine, it is more likely that the first commercial applications of these plant-derived chemicals will be in food processing which has less rigorous testing procedures than drugs manufacture. An example of this would be to coax plants to produce the starch-digesting enzyme α-amylase, which is used to make foods (including bread) and to clarify wines.

It is also possible to use gene transfer technology to re-direct the biosynthetic pathways of plants. By this means plants can be persuaded into producing more of the products they normally make, such as edible oils, waxes, lipids and starches, as well as things they would not normally produce. An interesting example of this is the genetic engineering of *Arabidopsis* to produce granules of polyhydroxybutyrate (PHB), a polyester used to make biodegradable plastic (a product discussed in detail later in this chapter). This product is normally obtained from the bacterium *Alcaligenes eutrophus*, which is genetically engineered to store granules of PHB. The plant can survive whilst storing PHB in the same way as the bacterium. Improvements still need to be made to the process to encourage the plant to concentrate the PHB in the small intracellular structures known as plasmids and to overcome the poor growth and seed production of the plants that yield most PHB. Notwithstanding these difficulties this third wave of plant biotechnology offers a great new opportunity, and potentially a means of overcoming the problem of surplus foodstuffs if plants can be encouraged to produce much more than just food and fibre.

THE EFFECT OF BIOTECHNOLOGY ON PROCESS TECHNOLOGY

Fermentation Technology

Biotechnology can be applied to various industrial manufacturing processes that use bacteria, yeasts, fungi, algae and plant and animal cells. The most commonly used of these process technologies is fermentation. The origins of fermentation technology go back a long way to the production of many traditional foods and beverages such as beer, cheese and wine. Modern biotechnology acknowledges the role that microorganisms can play in substituting for many costly, pollutive and inefficient chemical processes. Microorganisms can be used to produce many life-saving drugs such as penicillin by a simple extension of the domestically used fermentation process.

The process of commercial fermentation is essentially the same for all products: a large number of cells of uniform characteristics are grown under controlled conditions. At root, fermentation is the interaction of the microorganism with its source of nutrition. For commercial purposes, this process is carefully controlled to make it as efficient as possible and to obtain the optimum yield. It is also important that the process can be repeated with as little variation as possible in the quality of the end-product.

The receptacle for the fermentation process is usually known as a bioreactor or fermenter. Traditionally, this has been a tall, sterile column but recently experiments have been carried out to vary the design, in order to increase the rate of product formation and improve the quality of the product. Experiments have shown that stirring the contents of the bioreactor (with a paddle, for example) accelerates the reaction between the microorganism and its nutrient. There is some controversy over the optimum stirring rate, as too rapid agitation has proved to be counter-productive. It is vitally important that the reactor design and operation should ensure that aseptic conditions are main-

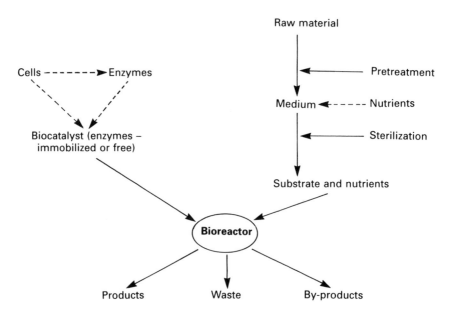

Fig. 3.2 Steps in bioprocessing (OTA, 1984).

tained since a fermentation batch can easily 'go off', thus wasting large amounts of the microorganism and its nutrients. This is a costly and not uncommon problem of biotechnology.

The application of biotechnology to industrial fermentation processes can bring about significant improvements. For example, the Brazilian bioethanol programme depends on a batch fermentation system; a continuous method of fermentation would speed up the process and save money. Continuous fermentation techniques are under study; the preferred approach relies on the retention of yeast cells in the bioreactor by separation and by continuous evaporation. Other areas of research include the use of genetic engineering to produce more efficient microorganisms that offer improved carbohydrate fermentation, resistance to high temperatures and high alcohol levels, higher speed of fermentation and higher yields. Novel fermentation techniques have been developed, such as fermentation under partial vacuum and recycling the yeast cells, and these have increased ethanol productivity by 10–12 times. Application of these biotechnological improvements to continuous production will make these processes increasingly economically attractive for the production of a substitute for fossil fuel (Smith, 1988).

Bioreaction Times

One of the fundamental difficulties with fermentation is to increase the efficiency of the process. According to Smith (1985) processes can be broadly considered to be either conversion cost-intensive or recovery cost-intensive. With conversion cost-intensive processes the volumetric productivity, Q_p (x kg of product m^{-3} hr^{-1}, volumetric production), is of major importance, while with recovery cost-intensive processes the product concentration, P (kg m^{-3}), is the main criterion for the minimization of cost. Put simply, in a typical biological fermenter there is a very low concentration of the reaction product. Whereas in a typical chemical process heat would be applied to accelerate the reaction, this is not possible in a biological reaction as it would kill the living organisms.

In most cases reactions occur between 20 and 65°C and at near-neutral pH conditions. Research has concentrated on optimizing the process, that is, minimizing the use of the raw material, maximizing the yield, and using energy efficiently. The reaction is tested to the limit of its physical parameters in order to discover the optimum conditions in which it should take place. The growth of the cell material in a bioreactor can be measured in terms of either dry or wet mass. There are two expressions used to describe the length of the process: one is doubling time (t_d), which refers to the period required for a doubling

Table 3.1. Average doubling times of cells/hours.

Bacteria	0.25–1.0
Yeast	1.15–2.0
Fungi	2.0–6.9
Plant cells	20–40

Table 3.2. Production data on fermentation products.

	Yield on substrate/%	Concentration/ $g\,l^{-1}$	Time/ hr	Productivity/ $g\,hr^{-1}$
Ethanol	45–50	110	0.5	3.67
Gluconic acid	90	15	0.5	0.39
Vitamin B_{12}	0.1	0.6	2	0.005

in mass of the desired product to occur; the other is generation time (g) which refers to the period of growth. Average doubling times for cells increase with cell size and complexity (see Table 3.1).

The same concept of doubling time can be applied to the products of fermentation, as well as to cells. These have a much longer doubling time (Table 3.2). Both the longer doubling time and the lower concentration are reflected in the cost of the product.

These doubling times have been assessed under laboratory conditions. But growth rarely occurs at an optimum rate in a commercial continuous batch cultivation. In practice, there are several distinct phases and different rates of growth (Figure 3.3). There is usually an initial *lag phase* (1) during which time no growth is apparent but substantial metabolic change is occurring as the microorganisms react with the nutrient medium. There then follows an *acceleration phase* (2) as the microorganisms begin to grow at an ever-increasing rate until *exponential growth* (3) is achieved when growth is unlimited. This phase is short-lived in most batch cultivations, as the reaction is soon limited by the availability of nutrients and the *deceleration phase* (4) begins. Finally the reaction reaches a *stationary phase* (5) when no further growth occurs as the nutrients have been exhausted. This is followed by a *decline phase* (6) and ultimately, if allowed to continue uncontrolled, the *death phase* when the rate of growth becomes negative. In most biotechnological batch processes conducted for commercial purposes, however, the reaction is stopped short of this phase.

There is an important relationship between the concentration of the nutrient and the rate of growth. A great deal of work was

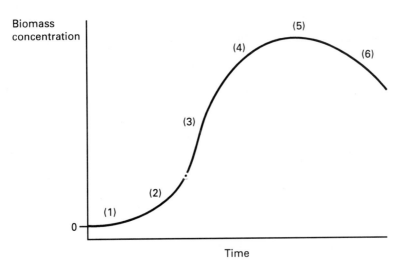

Fig. 3.3 Growth characteristics of a typical batch cultivation (Fiechter, 1981).

undertaken by Monod in the 1940s to develop a mathematical equation which describes this relationship. The original equation was expressed as:

$$u = u_{max}\left[\frac{S}{K_S + S}\right]$$

In this case, S is the concentration of the substrate in the medium in which the reaction occurs, u_{max} is the maximum specific growth rate of the organism and K_S represents a saturation constant, the substrate concentration at which $u = u_{max}/2$. Thus exponential growth can occur at specific growth rates having any value between zero and u_{max}. This is particularly difficult to achieve with continuous batch fermentation. Strenuous research efforts have been devoted to maintaining the right level of nutrient concentration in a continuous process. To achieve this the level of nutrients has to be maintained whilst a proportion of the desired product is removed. It becomes very important to maintain the correct pH balance in a continuous culture as different stages of the process require different pH conditions. A balance has to be maintained between the inflow of the nutrients and the inoculum (f) (culture inoculated with the bacterium) and the outflow of the completed culture (v). The rate of dilution (d) of the contents of the bioreactor can therefore be described as $d = f/v$. If dilution exceeds the specific rate of growth, then 'wash out' occurs and the continuous process breaks down unless the situation is swiftly corrected. In an efficient system of continuous fermentation, the quantity of cells produced per unit of time should be only fractionally less than the optimum rate of output and substrate concentration.

Bioreactor Design

Continuous systems require very stringent safeguards against contamination. Before commencing the reaction the bioreactor and the pipework have to be sterilized, usually with high-pressure steam. Air which enters the bioreactor has to pass through a filter, usually of sterilized glass wool. The nutrient medium itself has to be kept free of contaminants as it is obviously an ideal breeding ground for a whole host of undesired microorganisms. The cost of having to abandon a contaminated batch can be substantial, especially for high-value products such as antibiotics; for such products a smaller bioreactor may be used in order to limit the loss. Bioreactors are usually made of highly polished stainless steel, which can withstand high-pressure steam and resist corrosion. There are numerous designs for bioreactors but the most common are columns or towers of varying height. The reason for this is simply the law of gravity, whereby the nutrient and the inoculum enter at the top of the bioreactor and the reaction product is drawn off ('harvested') at the bottom.

Bioreactors tend to yield more if stirred, either by mechanical means or by putting a gas through the medium. The basic design of the continuously stirred tank reactor goes back to the 1940s, when the industrial production of penicillin commenced. The mixing and the dispersion of the ingredients of the reaction can entail a high energy cost. This is one of the main incentives for switching to the agitation of the mixture by the throughput of sterile air; but care has to be taken not to damage the organisms, and the size of the gas bubbles is therefore often controlled. This can present problems when the mixture is particularly viscous.

As stated earlier, one of the difficulties with bioreactions is the low concentration of the product. One of the consequences of this is that large quantities of waste water have to be disposed of. Depending on the concentration of the product, the waste water may have to be treated before it can be discharged into the public water system. One efficient way of dealing with this is to convert the remaining organic material into carbon dioxide or methane, so-called 'biogas', by anaerobic or aerobic reactions. Methane so derived may be recycled into the energy system of the whole process. This is an important technical development which can have a positive economic bearing on the costs of production. It has been particularly successful in contributing to a positive energy balance in the production of bioethanol; for example, in the pilot scheme for ethanol and biogas in Ochsenfurt (Kammerer, 1984) the ratio of energy output:input was increased from 6:1 to 8.6:1 when biogas was produced and used.

Scaling up can cause problems in design and control of the process. To give an idea of the different scales of production: a laboratory-scale plant consists of a fermenter of 5–10 litres, a pilot-scale fermenter has a capacity of 100–10,000 litres, and a full-scale industrial fermenter one of 40,000–400,000 litres. When scaling up there are particular problems with removal of heat, availability and transfer of oxygen, mixing the medium and production of toxins. A high level of capital investment is required to manage these problems, but the incentive to safeguard against them is high because the cost of aborting a batch increases with the scale of the fermenter.

Bioreactor Nutrition

The nutrient medium usually consists of a source of carbon since living organisms need this as a source of energy. An available nitrogen source is also required for life and growth to be sustained. Sources of both elements are listed in Table 3.3. For most biotechnological reactions the cheapest sources of these are sought in order to minimize the costs of production.

Obviously there are wide variations in the price per unit of carbon or nitrogen amongst these nutrient sources (this is discussed in Chapter 4). The degree of purity of the source also has an important impact on the process itself and upon the quality of the waste water. For example, molasses (51% sugar content and 87% purity (ash remaining after water and sugar)), used as a source of carbon as well as of nitrogen, is a considerably less pure carbon source than sucrose, and creates considerable effluent problems which have to be offset against the fact that

Table 3.3. Carbon and nitrogen sources used in biotechnology.

Sources of carbon	Sources of nitrogen/%N by mass
Glucose	Barley (1.5–2.0)
Lactose	Beet molasses (1.5–2.0)
Starch	Corn steep liquor (4.5)
Sucrose	Soyabean meal (8.0)
	Groundnut meal (8.0)

Source: Rhodes and Fletcher (1966)

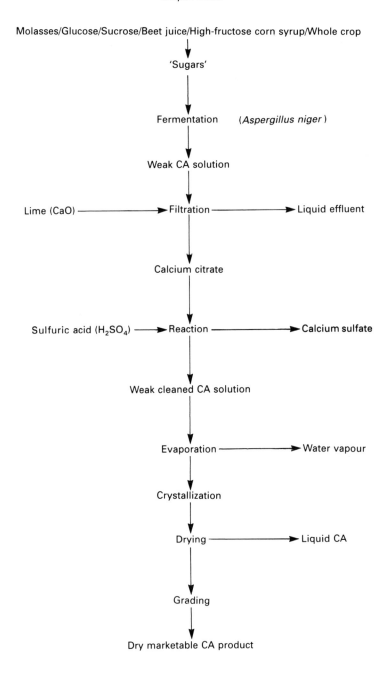

Fig. 3.4. Schematic representation of the citric acid (CA) production process.
Source: J. and E. Sturge, Selby, Yorkshire, pers. comm.

it is cheaper than sucrose. The cost of the carbon and nitrogen sources can represent as much as 60–80% of the variable costs (Hacking, 1986). Trace metals, vitamins and some amino acids and fatty acids also have to be added to the nutrient medium to ensure optimum growth. Fig. 3.4, a schematic representation of citric acid manufacture, indicates the complexity of the process.

Another interesting feature of bioreactor nutrition is the specificity of the preferred diet of the microorganisms. This is well illustrated by the dietary habits of the mould *Aspergillus niger*, used in citric acid manufacture. Once used to a diet of a less pure substrate, such as molasses, the mould performs less well on the purer sugars such as glucose and sucrose. The reverse is also true if the mould is reared on the purer substrate from the start. The reasons for this are not well understood but it is an observed fact that the microorganisms appear to develop a 'taste' for a particular substrate. This undermines the possibilities for technical interchangeability, and sometimes prevents the cheapest substrate from being used.

Another nutritional possibility being investigated is solid substrate fermentation, in which microorganisms are grown on solid material with virtually no free water. Biological activity ceases when the moisture content of the substrate falls below 12%. However, it is possible to sustain biological activity close to this level. The area of greatest potential would appear to be the use of compounds that are insoluble or virtually insoluble in water, such as wood, cereal grains, seeds of legumes and wheat bran. They tend to be cheaper to use than the refined products made from them. So it is anticipated that solid substrate fermentation of lignocellulose could serve well for the production of ethanol and methane. The only energy-consuming part of the operation is the reduction of the substrate into particles of optimum size for the physical parameters of the fermentation. The main benefits of solid substrate fermentation are that cheaper substrates can be used, and that energy costs are less. Set against this, the reaction is more difficult to control, particularly since substantial amounts of heat are produced from the metabolic reaction, and because the growth of the microorganisms tends to be slower than in liquid fermentation.

Enzyme Technology

The agents responsible for fermentation are known as *enzymes*. They are highly specific biocatalysts which cause biological reactions to take place and can enormously increase the speed of a reaction. Scientific understanding of how enzymes work dates back to the late 19th century, which is comparatively recent compared to the length of time people have been happily using them. The advent of molecular biology has accelerated the rate of application of enzymes to commercial ends. However, although thousands of enzymes have been identified, only about 20 are used to any significant extent in industry. These are the amylases, proteases, pectinases and cellulases that have found widespread application amongst food and drink manufacturers.

There are several specific obstacles to overcome if an even wider application of enzymes is to be achieved. All enzymes need mild physiological conditions to operate in, and are sensitive to alterations in temperature or the pH value of their surroundings. They are easily affected by microbial contamination and difficult to recover from the fermentation solution. A great deal of attention has been paid to the latter problem, and techniques have been developed for immobilizing the enzymes in the fermentation solution. There are considerable economic benefits if the biocatalyst can be retained in a process of continuous fermentation.

Many immobilization techniques have been developed but the most common methods are:

- covalent attachment to the surface of a water-insoluble material, such as porous glass, ceramics, nylon, alumina or cellulose (however, the formation of covalent bonds can render the enzyme partially or wholly inactive);
- entrapment in gels which have the advantage of being particularly mild and unlikely to damage the enzyme activity (however, there can be loss of enzyme through the gel pores);
- encapsulation within partially permeable membranes such as collodion, polystyrene and nylon, which are impermeable to the enzyme but allow the product through;
- adsorption on solid surfaces of inorganic materials such as clays, glass, alumina and carbon (but the process needs to be reversible);
- cross-linking with multifunctional agents, creating water-insoluble aggregates.

The two best-known and most successful examples of immobilized enzymes used industrially are *glucose isomerase* (used in the food industry) and *penicillin acylase* (in the pharmaceutical industry). The development of a technique for the immobilization of glucose isomerase by entrapment in polyacrylamide matrices created a market opportunity to undercut the traditional sugar industry with a cheaper liquid sweetener, produced by the enzymically catalysed partial conversion of glucose to fructose. In practice, the enzyme can use a cheap source of starch and react at a high level of concentration and at a comparatively high temperature (60°C) and high pH value.

The second most important industrial application of enzyme

immobilization technology is to produce penicillin from 6-aminopenicillinic acid (6APA), using the enzyme penicillin acylase. The process takes place in a fixed-bed column bioreactor; to give an idea of the efficiency of the immobilization technique, only 30 tonnes of the enzyme are required to produce 3500 tonnes of 6APA.

One of the major obstacles to the extension of enzyme technology has been the legislative restrictions placed upon their use. Whilst pure enzymes are rarely toxic, they can form hazardous compounds such as mycotoxins or antibiotics. Furthermore, it is possible for the fermentation medium to become contaminated with 'foreign' microbes. For these reasons, there are strict safety measures in force on the use of enzymes in food production. The cost of safety testing tends to act as a deterrent to industrial companies to seek approval for new enzymes, with the result that the sources of industrial microbial enzymes are remarkably limited. The present knowledge of enzyme technology is based on the disciplines of molecular biology, enzymology and physical chemistry; but the next generation of biotechnological applications of enzymes will draw on a wider range of expertise from biophysics, electrochemistry, biochemical engineering, organic and polymer chemistry, microbiology, biochemistry and genetic engineering. The objectives of bringing together this wider range of disciplines will be to produce multienzyme systems which are capable of advanced functions such as simulating the work of vital human organs.

Summary

To conclude this section on bioprocess technology it is useful to recall the advantages and disadvantages of using bioprocesses over conventional technology as listed in Table 3.4.

Table 3.4. The advantages and disadvantages of bioprocess technology.

Advantages	Disadvantages
Milder reaction conditions	Susceptibility to contamination
Use of renewable resources	Low reaction concentrations
Less expensive raw materials	Slow reaction rates, cannot be accelerated by heating
Reduced environmental risk	
Simple manufacturing base	Large amounts of waste water
Lower capital investment	Complex product mixtures requiring separation and purification
Reduced energy consumption	
Greater specificity of catalytic reaction	Genetic instability
	Need to contain modified organisms

There are some cases where bioprocesses have to be used industrially because they are the only practical way of producing desired products such as antibiotics and proteins. Another reason why bioprocessing may be chosen is that the sequential reactions executed by microorganisms may be impossible to imitate by conventional chemical means. In most cases the industrialist has to weigh up the advantages and disadvantages of the two methods before making an investment. However, biotechnology is bringing improvements all the time to bioprocessing; with a combination of genetic engineering and process design, bioprocesses are becoming more efficient and economical and therefore more competitive.

Considerable effort has gone into developing cultured plant cells as a system for producing valuable chemicals. For example, suspension cultures of the rosy periwinkle, *Catharanthus roseus*, have been grown in bioreactors to produce indole alkaloids of medicinal value (Facchini and Dicosmo, 1990; Jardin *et al.*, 1991). Bioprocessing with plant, mammalian and insect cells will extend the range of the technology to an increasing number of new products.

THE IMPACT OF BIOTECHNOLOGY ON NON-FOOD PRODUCTS

The effects of biotechnology on product formation are threefold: biotechnology creates *novel* products, *substitute* products and conventional products with *new characteristics*.

A good illustration of a product which embraces all three of these aspects is the bioplastic developed by ZENECA (then ICI), known as polyhydroxybutyrate (PHB). The technology for the production of this bioplastic came from the expertise which ICI developed using bacteria to convert methanol to single-cell protein. PHB is also produced from a controlled bacterial fermentation. A microorganism, *Alcaligenes eutrophus*, was discovered which produces bioplastic granules within its cell wall. Normally the bacterium is rod-shaped, but when filled with bioplastic it swells. The nutrient it requires for this process is a carbohydrate substrate, such as sugar or starch, together with a proportion of propionic acid, about 7–12% of the nutrient medium. In order to harvest the bioplastic the cell wall has to be ruptured without damaging the plastic. This is a delicate process since the plastic is destroyed at temperatures above 70°C and at pH higher than 8. In the past a solvent was used for this purpose but it proved too costly. Therefore research efforts are now being directed to cheaper methods of recovering the product. It has recently been demonstrated (as mentioned earlier in this chapter) that PHB can be produced in transgenic plants. This offers the possibility of producing large amounts of PHB at low cost, as plants provide their own carbon source via atmospheric carbon dioxide and photosynthesis.

Once the plastic has been recovered, it is concentrated in centrifuges and then crystallized. The size of the crystals has an important bearing upon the versatility of the plastic. If the crystals are allowed to become too large the plastic produced from them becomes too brittle to be of practical use. One way ICI have found of overcoming this problem is to add another plastic known as polyhydroxyvalerate (PHV) which gives the bioplastic mix greater flexibility. Once the plastic has been recovered, it can be extruded and moulded into any of the shapes of traditional plastic artefacts; it can also be spun into a fibre or produced in film form.

PHB resembles conventional oil-derived plastic in many ways but it also has novel attributes: it is biodegradable and biocompatible. Artefacts made from PHB degrade by surface erosion in the environment. This means that PHB, unlike its oil-derived equivalents, can be buried in a rubbish tip and simply break down into harmless compounds – carbon dioxide and water – with no damage to the environment. The speed at which the articles break down can be controlled; this is necessary, because it would be impractical for bioplastic buckets and drainpipes to break down prematurely. The speed of degradation is affected by the environmental temperature, porosity, the microbial population of the soil, the section thickness of the plastic and its surface texture. So, for example, a plastic bottle when buried will degrade completely in nine months whereas bioplastic film of 100 microns (0.01 mm) thickness will degrade completely in 21 days. This property makes PHB an interesting possibility for a mulch for use in agriculture or horticulture.

PHB is non-toxic, and experiments have shown that it is compatible with living tissue; it does not cause inflammation when inserted in the body. This fact has opened up a whole range of medical applications for the bioplastic such as in wound care and also in paramedical areas such as colostomy bags and nappies. It can also be produced in such a way as to break down inside the body if desired. The product of degradation is 3-hydroxybutyric acid, a chemical that is found in normal mammalian tissue in concentrations of between 3 and 10 mg per 100 cm^3 of blood.

This bioplastic thus has a wide range of applications in agriculture and health care and for domestic use. It is a novel product produced by completely new technological means; it can substitute for most conventional uses of oil-derived plastics (leaving aside economic considerations for the present); and it adds new characteristics to plastic of complete biodegradability and biocompatibility.

Conclusion

This chapter has shown that biotechnology, an integrated amalgam of respected scientific disciplines, has opened up a whole new range of opportunities for agriculture affecting both its input side and its outputs. It has to be remembered that this technology is still in its infant stages and many of the teething problems have been discussed, particularly with respect to scaling up and product recovery. These technical difficulties have engaged research scientists and commercial companies in the active search for improvements, motivated by the commercial potential they perceive. How correct their perception may be will have to be assessed in conjunction with the rigorous economic and political analysis that will follow. However, one thing is certain: the speed of technological breakthrough is increasing in this new area and vast resources have been attracted to this end from the private sector which should ensure that the momentum is maintained. The motives of the technologists may differ from those of politicians, economists and farmers but at least they offer the latter an increasing range of technological possibilities to choose from. One could go so far as to say that the whole development is technology-driven, except that in the final application the other interested parties have a controlling power – and rightly so, for technology alone may not produce the result which is in the best interest of society as a whole.

References

Benbrook, C. *et al.* (1986) Engineering crops to resist herbicides. *Technology Review* 89 Nov/Dec Issue.

Facchini, P.J. and Dicosmo, F. (1990) Immobilisation of cultured *Catharanthus roseus* cells using a fibreglass substratum. *Applied Microbiology and Biotechnology* 33(1), 36–42.

Fiechter, A. (1981) Batch and continuous culture of microbial, plant and animal cells. *Biotechnology*, Vol. 1. Verlag Chemie, Basel, pp. 453–505.

Fraley, R. (1988) Genetic improvements in agriculturally important crops. In: Fraley, R., Frey, N.M. and Schell, J. (eds) *Progress and Issues*. Cold Spring Harbor Laboratory, New York, 116 pp.

Franks, T. and Birch, R.G. (1991) Microprojectile techniques for direct gene transfer into intact plant cells. In: Murray, D.R. (ed.) *Advanced Methods in Plant Breeding and Biotechnology.* CAB International, Wallingford, pp. 103–127.

Gatehouse, A.M.R., Boulter, D. and Hilder, V.A. (1992) Potential of plant-derived genes in the genetic manipulation of crops for insect resistance. In: Gatehouse, A.M.R., Hilder, V.A. and Boulter, D. (eds) *Plant Genetic Manipulation*

for Crop Protection. CAB International, Wallingford, pp. 155–182.
Grant, J.E., Dommisse, E.M., Christey, M.C. and Conner, A.J. (1991) Gene transfer to plants using *Agrobacterium*. In: Murray, D.R. (ed.) *Advanced Methods in Plant Breeding and Biotechnology.* CAB International, Wallingford, pp. 50–73.
Hacking, A.E. (1986) *Economic Aspects of Biotechnology.* Cambridge University Press, Cambridge.
Jardin, B., Tom, R., Chavarie, C., Rho, D. and Archambault, J. (1991) Stimulated indole alkaloid release from *Catharanthus roseus* immobilised cultures. *Journal of Biotechnology* 21(1–2), 43–62.
Kammerer, F. (1984) *Pilotprojekte für Bioethanol und Biogas in Ochsenfurt.* CIBE.
Kidd, G.H. (1985) The new plant genetics – restructuring the global seed industry. In: *The World Biotech Report,* Online Publications, London.
Marshall, G. (ed.) (1989) *Flax: Breeding and Utilisation.* Kluwer Academic Publishers, Dordrecht.
Moffat, A.S. (1992) High tech plants promise bumper crop of new products. *Science,* 256 (8 May 1992) 770–771.
Mullineaux, P. (1992) Genetically-engineered plants for herbicide resistance. In: Gatehouse, A.M.R., Hilder, V.A. and Boulter, D. (eds) *Plant Genetic Manipulation for Crop Protection.* CAB International, Wallingford, pp. 75–108.
OECD (1989) *Biotechnology, Economic and Wider Impacts.* OECD, Paris.
Office of Technology Assessment (1989) *New Developments in Biotechnology: Patenting Life.* Special Report, OTA-BA-307 US Government Printing Office, Washington DC.
OTA (1984) *Commercial Biotechnology.* Pergamon Press, Washington, DC.
Owen, T.R. (1986) Nitrates in drinking water. *Fertilizer Research Review* 10. Dr W. Jonk Publishers, Dordrecht.
Persley, G.J. (1990) *Beyond Mendel's Garden: Biotechnology in the Service of World Agriculture.* Biotechnology in Agriculture Series No. 1, CAB International, Wallingford.
Rathus, C. and Birch, R.G. (1991) Electroporation for direct gene transfer into protoplasts. In: Murray, D.R. (ed.) *Advanced Methods in Plant Breeding and Biotechnology.* CAB International, Wallingford, pp. 74–102.
Rhodes, A. and Fletcher, D.L. (1966) *Principles of Industrial Microbiology.* Pergamon, Oxford.
Smith, J.E. (1985) *Biotechnology Principles.* Van Nostrand Reinhold, Maidenhead, UK.
Smith, J.E. (1988) *Biotechnology,* 2nd edn. Edward Arnold, London.

4

THE COST EQUATION:
THE ECONOMICS OF NON-FOOD USE

Many of the outlets for the non-food use of agricultural raw materials (ARMs) are not yet commercially viable. This is not necessarily the same as economically viable. For example, commercial viability may be hampered by farm policies which unduly increase the market price of agricultural products or by energy policies which fail to recognize the social costs of burning fossil fuels.

In several cases the ARMs have to compete with cheaper non-agricultural raw materials (NARMs), often derived from oil. Unless there is a real change in the relative price of ARMs compared to NARMs, the opportunities for non-food uses are unlikely to be developed. This chapter examines the development of prices of the competing raw materials: oil-derived and agricultural, in an effort to establish the true competitive position between the two. It also identifies the shortcomings of pricing and forecasting methods. However, there is an underlying trend of the cheapening of ARMs relative to oil. Even before the 1990 Gulf crisis, ARMs had cheapened to the extent that it was possible to buy four times as much cereal for the equivalent value of a barrel of oil as 20 years ago.

There is wide variation in the economics of non-food products ranging from speciality chemicals, such as antibiotics, where the price of the raw material is insignificant compared to the value of the final product, through to bulk chemicals of low value where raw material costs are crucial. Two case studies are given to illustrate this point: one is the energy use of ARMs to produce bioethanol (as a petrol additive) for which the ARMs have to compete with products derived from oil. This product is made in large quantity in Brazil and the USA, but in both countries it has to be subsidized. It is not yet produced in any significant quantity in the European Community and this chapter analyses the possibility of that situation changing. The second case study is of a biodegradable plastic made from ARMs but which still has to compete with oil-derived alternatives. There is a wide range of applications for this bioplastic from high-value medical uses such as in colostomy bags to medium-value uses as biodegradable packaging and wrappings.

The chapter concludes by examining alternative methods used to encourage non-food outlets and those specifically envisaged in the European Community under the so-called MacSharry reforms which were adopted in May 1992. These included measures allowing land taken out of food production to be used for the production of cereals for non-food purposes.

COMPETITION BETWEEN ARMs AND NARMs

For the purposes of examining the price relationship between NARMs and ARMs, oil has been chosen to represent the former and cereals the latter, because these are the two most widely used raw materials in each case.

The Development of Oil Prices

The price of a sample of oil depends on the region of the world from which it has been extracted and also on the quality of the oil. A broad distinction may be drawn between heavy and light oils. The 'lightness' of an oil basically describes the capacity to make petrol and ultimately gas from it. Oil pricing is heavily influenced by the official selling price of the oil-producing countries' association, OPEC. For the purposes of this chapter the Brent spot price has been chosen. A spot price is an immediate cash price derived from the idea of 'on the spot' sale as opposed to a price for forward delivery. The Brent price is considered the leading indicator of West European oil production. Brent crude oil is regarded as heavy and is given a rating of 37 in the American Petroleum Index (as compared with Arab light oil at 34). A better

Table 4.1. An illustration of crude oil composition.

	Arab (heavy)	Brent
Liquid petroleum gas (LPG)	2%	2%
Petrol	13%	23%
Kerosene	10%	13%
Diesel	18%	22%
Fuel oil	56%	40%

Source: UK Department of Energy, pers. comm.

Table 4.2. Brent dollar oil prices.

Year	$/bl^{-1} nominal	World GDP deflator	$/bl^{-1} 1985 prices[3]	$/bl^{-1} 1991 prices[3]
1970	2.29	20.40	11.23	23.29
1971	2.20	21.80	10.09	20.94
1972	2.32	23.30	9.96	20.66
1973	2.73	25.60	10.66	22.13
1974	9.74	29.60	32.91	68.28
1975	11.32	33.40	33.89	70.33
1976	12.80	37.10	34.50	71.59
1977	13.92	40.90	34.03	70.62
1978	14.02	45.10	31.08	64.50
1979	31.81	50.80	62.22	129.12
1980	36.83	58.20	63.28	131.31
1981	35.93	65.90	54.52	113.13
1982	32.97	73.80	44.67	92.70
1983	29.55	82.80	35.69	74.05
1984	28.66	91.30	31.39	65.14
1985	27.51	100.00	27.51	57.08
1986	14.38	107.60	13.38	27.73
1987	18.43	118.00	15.62	32.41
1988	14.96	133.50	11.21	23.25
1989	18.20	156.20	11.65	24.18
1990	23.81	185.60	12.83	26.62
1991	20.05	207.50[1]	9.66	20.05
1992	19.37[2]	232.29[2]	8.34	17.30

Source: British Petroleum (1992)
[1] World GDP figure is an estimate, 1991.
[2] World GDP figure and oil price are estimates, 1992.
[3] To calculate real price from nominal price and GDP:

$$\frac{\text{GDP deflator in 1970}}{\text{GDP deflator in 1985}} = \frac{\text{Nominal price in 1970}}{\text{Real (1985) price in 1970}}$$

$$\text{Real (1985) price in 1970} = \frac{\text{GDP deflator in 1985}}{\text{GDP deflator in 1970}} \times \text{Nominal price in 1970}$$

$$= \frac{100 \times 2.29}{20.40}$$

$$= 11.23$$

understanding of the value of crude oil can be obtained by comparing the variation in the elements derived from different types of oil. Table 4.1 illustrates what can be produced from one barrel of the crude oil. The percentages shown are not immutable. The eventual breakdown of these oil types into individual products will depend on the refining

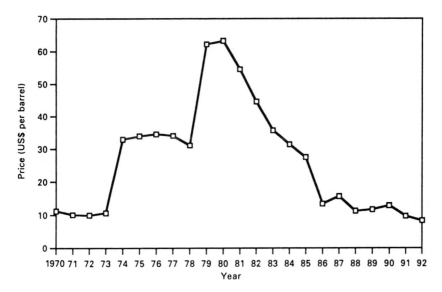

Fig. 4.1 Brent dollar oil price variations (1985 prices).

processes available and the varying price for each oil product.

There is another advantage of using Brent figures, in that the oil is produced in roughly the same global area as the ARMs against which its price development will be measured, and the series of figures are collected in a consistent and reliable manner. This price series did not commence until the discovery of North Sea oil in 1970. It is a price for crude oil and is quoted in US dollars.

The nominal oil price figures from 1970 to 1975 are taken from the *Monthly Review of External Trade Statistics* produced by the then UK Department of Energy – now part of the Department of Trade and Industry. The nominal figures for 1976 to 1991 come from the *BP Statistical Review of World Energy* (June 1992). The 1992 figure is a provisional average of the average daily one-month prices. The world GDP figures come from a GDP deflator series in the International Monetary Fund's *Statistics Yearbook*. The 1991 and 1992 figures were assumed to increase in line with world consumer prices, a series for which is given in the IMF *International Financial Statistics* (monthly) November 1992. These data are presented in graphic form in Fig. 4.1 in order to demonstrate the peaking and troughing of oil prices.

It is clear that there have been two sharp increases, so-called oil price shocks, and a price collapse in world oil prices since 1970. Throughout the 1960s the price of oil had remained relatively stable, ranging from US$1.21 to US$1.63 per barrel. A higher plateau in prices was reached after the October War in 1973, which was followed by the Arab oil embargo and a hike in OPEC oil prices. OPEC became a dominant force in world oil pricing and prices were sustained in the $10–14 range from 1974 to 1978. Another plateau was reached and sustained as a result of the Iranian Revolution and the panic buying that ensued after the interruption of oil supplies from Iran, which at the time was the world's second largest oil exporter. Spot prices reached $30–40 per barrel, at which level they remained throughout 1982, helped by the intensification of the Iran–Iraq war. By the mid-1980s the price increases had led to lower demand and higher supply. These developments caused prices to move down to a lower plateau of $27–30 per barrel for the period 1982–1985. Then in 1985–1986 OPEC shifted to a market share strategy and prices collapsed to $14 per barrel. The response of OPEC to the price collapse was to limit production through the adoption of quotas designed to sustain prices around the $18 level. This plateau was broadly maintained until the Gulf crisis of 1990, which had an immediate and dramatic effect upon world oil prices. During the Gulf crisis some governments intervened to fix oil prices internally in order to mitigate the effects of the sudden increase in prices for industry. This increase was not sustained and prices returned to their $17–18 plateau. Subsequent military tension in the Gulf tended to produce a similar pattern of short-term price increases which returned to 'normal' when the tension abated.

There has certainly been an underlying real increase in the price of oil since 1970. This may in part reflect the fact that oil is a finite resource and thus its price may be expected to rise as supplies run out. In fact, new sources of oil continue to be found through exploration; however, they are not always as easy to extract as were the first discoveries. Furthermore, 70% of the world's oil reserves are now concentrated in the Middle East – a region not noted for political stability, as the Gulf crisis demonstrated. Therefore the price of oil also reflects in part the strategic risk of supplies being disrupted. This is not to say these factors are fully reflected. Neither does the price of oil reflect the cost of the environmental damage caused by burning fossil resources. It is in their combustion that harmful gases are released into the Earth's atmosphere which profoundly disturb the ecological balance. The latter is a relatively recent and still controversial scientific discovery and thus it is probably true to say that it is not yet reflected in the oil price (CEA, 1988). Such costs are very difficult to quantify and are more likely to be reflected in the taxation element of fuel pricing as each country assesses the importance it attaches to this phenomenon. An illustration of this is the proposal by the European Commission to introduce a carbon tax on fossil fuels, which would increase progressively to $10 per barrel. There is some reason to believe that commercial interests should take account of future levels of oil scarcity by increasing prices now. However, there is no commercial reason to believe that commercial market forces will take account of the social cost of global warming in the pricing of fossil fuels – hence the need for governments to intervene via the instrument of fuel taxation.

Oil Price Forecasts

Efforts are made to forecast the price of oil but they are notoriously difficult to get right. The *Oil and Gas Journal*, published in Oklahoma, regularly reviews a range of forecasts produced by both those within the industry and independent analysts. Because of their location in the USA, they use the West Texas Intermediate price (WTI). WTI prices are normally one or two dollars higher than Brent prices. Table 4.3 shows the results of their survey of forecasts made in 1992. Their range of forecasts for the year 2010 would produce an average figure of US$47.05 per barrel, which in 1991 US dollars would mean a price of US$19.33 per barrel.

The International Energy Agency (IEA), part of the Organization

Table 4.3. West Texas Intermediate spot price forecasts/$ per barrel.

Source	Date	1992	1995	2000	2005	2010
County Natwest	9/92	20.95	22.00	–	–	–
Petrie & Co	10/92	20.00	24.00	30.63	33.77	35.46
Canadian Oil	4/92	20.00	21.85	26.46	32.19	39.17
Analyst D	8/92	20.50	23.73	30.29	33.39	35.06
Analyst E	4/92	20.80	26.90	–	–	–
Analyst J	4/92	22.12	23.84	24.70	29.58	51.11
Ind oil co A	9/92	20.92	20.90	29.93	42.85	61.36
Ind oil co C	4/92	19.50	26.00	33.75	–	–
Major oil co D	11/92	20.50	22.66	28.98	38.20	51.77
Major oil co H	9/92	20.59	26.25	31.50	35.75	40.00
Major oil co I	2/92	20.00	23.06	29.42	38.69	53.01
Major oil co K	8/92	21.00	20.70	22.99	27.00	–
Major oil co M	11/92	21.00	25.36	32.37	35.69	37.47
Major oil co O	5/92	20.00	23.60	31.60	42.25	56.55
Range						
High		22.12	26.90	33.75	42.85	61.36
Low		19.50	20.70	22.99	27.00	35.06
Average		20.47	23.61	29.55	36.29	47.05

Source: Oil and Gas Journal (1992)

Table 4.4. Oil price forecasts 1995–2010/1989 US dollars per barrel.

	1995	2000	2010
Most likely (= Sc.1)	17.5	20.0	30.0
Increased economic growth but little conservation (= Sc.2)	20.0	26.5	40.0
Growth + conservation (= Sc.3)	20.0	25.0	20.0

Source: DG XVII European Commission (1989)

for Economic Development and Co-operation, is internationally recognized as an authority on oil price forecasts. In its *Review of the Energy Policies of the OECD Countries* (1991), the IEA give two price scenarios: the first is a level price scenario where the price of oil does not exceed US$20.8 by the year 2005; in the second, the oil price rises to $35.5 by that date.

The European Commission also produced a range of oil price scenarios in 1989. The Community presently imports 70% of its oil but the Commission fears this could rise to 85% by the end of the century. The European Commission does have to undertake a certain amount of energy planning to ensure that the Community's energy demands are broadly in line with supply. In making such an assessment the Commission assumes there will be no significant variation from current trends. In making its projections in July 1989 (Table 4.4) the Energy Directorate (DG XVII) used three different scenarios:

Scenario 1: a continuation of present trends in economic growth and efficiency of oil use with no application of any specific energy-conservation methods; the oil price is estimated under this scenario to increase by 70% above its 1989 level by the year 2010.

Scenario 2: higher than expected economic growth and oil in shorter supply than currently estimated by the first decade of next century; the oil price is estimated to increase by 122% by the year 2010.

Scenario 3: higher growth than currently expected but combined with the application of the best available energy-conservation methods and no change in oil supply expectations; the oil price would only increase by 11% by the year 2010.

The European Commission considers Scenario 1 the most likely, i.e. that the price of oil will not exceed US$20 (1989 dollars) per barrel by the end of the 1990s. Under the worst possible scenario it could rise to $26.5 per barrel by the year 2000, and to $40 per barrel by 2010. Therefore it is against this range of oil values – i.e. from $20 to $40 per barrel – that the viability of substituting oil with ARMs can be judged.

Oil companies naturally tend to produce their own oil price forecasts, which are often bullish. The British Petroleum company forecasts the price of Brent to be $25 (1991 dollars) per barrel for both 2000 and 2010.

The Development of Agricultural Prices

The pricing of ARMs has certain features in common with that of NARMs, the most important of which is intervention or regulation of prices. Just as there is an official selling price amongst oil-producing countries, so most major agricultural producing nations regulate their domestic agricultural prices to some extent. In the European Community (EC), a common price level applies across all member states. However, this price level is often unrelated to world market price levels or to the domestic price-fixing arrangements of other agricultural producers such as the USA and Japan. There is a world market price for the main agricultural products; quantities of produce surplus to domestic requirement usually receive this price, the price differential (if any) being met by an export subsidy. The percentage of each commodity traded freely on the world market varies considerably, but one good reason for choosing cereals to provide a benchmark for the development of the price of ARMs is that they are widely produced and traded throughout the world. Another good reason is that starch

Table 4.5. The development of wheat prices[1] 1970–1990/US$ per metric tonne, 1980 prices.

Year	Nominal	Real	Year	Nominal	Real
1970	57.0	163.8	1980	168.3	168.3
1971	62.1	169.2	1981	154.6	153.8
1972	69.1	172.8	1982	132.6	133.8
1973	136.8	294.8	1983	137.3	142.1
1974	178.0	315.0	1984	140.2	147.7
1975	138.4	220.4	1985	128.7	134.2
1976	122.7	192.6	1986	118.4	104.4
1977	95.5	136.4	1987	112.1	90.8
1978	124.9	155.2	1988	140.7	106.2
1979	156.3	171.4	1989	161.3	122.7
			1990	129.1	92.7

Source: World Bank (1991)
[1] US soft red winter wheat f.o.b. Atlantic ports.

derived from cereals is a key raw material used by the chemical industry for non-food purposes.

The price series used for cereals (Table 4.5) is taken from World Bank data; it shows the relative cheapening of cereals over the last 20 years.

The Underlying Economic Trend

At present, the interdependence between oil and ARMs is largely in one direction: oil products are used in the production of ARMs but the reverse is not yet true to any great extent. Fertilizers, herbicides, fungicides, pesticides and fuel are all products of oil without which agriculture would be severely handicapped. The main reason for the lack of ARMs used to substitute oil-based products is that they are not commercially viable. Their cost is too high to attract oil producers to their use. However, the economic situation is not static. The underlying economic trend is of the relative cheapening of agricultural raw materials compared with oil. Over the last 20 years ARMs have been cheapened in relation to oil to the extent that it is possible to buy four times as much wheat with 1 tonne of oil in 1990 as in 1967.

This is illustrated in Table 4.6, where the development of feed wheat prices is compared to the development of crude oil prices since 1967. A calculation is then made of the volume of feed wheat which can be purchased with the equivalent value of a barrel of crude oil. The distorting factor of currency fluctuations is reduced by quoting both sets of prices in commercial ecus. Although the table compares the market price of oil with a supported price for wheat, there has been some convergence between the market price and support price for wheat in the EC over this period. This trend has received fresh impetus from the May 1992 decisions taken in connection with the GATT trade talks, which are designed to reduce the level of market support.

As can be seen from Table 4.6, during the mid-1980s when oil prices averaged over 30 ecu per tonne, the quantity of wheat which could be purchased with the value of one barrel reached eight times the 1967 level. The data in Table 4.6 are shown in graphic form in Fig. 4.2, which again demonstrates the competitiveness of ARMs.

This is not to say that the energy value of one tonne of wheat is equal to that of one barrel of oil. Indeed one tonne of alcohol derived from wheat only has one third of the energy value of one barrel of oil. It takes 280 kg of wheat to make one hectolitre (hl) of alcohol. In addition, the two products require different amounts of energy to process them, and the fuels perform differently in engines. 100% alcohol fuel has only two thirds of the energy value of petrol. This is a purely

Table 4.6. Prices for feed wheat and crude oil 1967–1990.

Date	Feed wheat[1]	Crude oil[2]	GDP deflator[3]	kg wheat[4]
1967	95.0	2.1	25.1	22
1968	98.3	2.2	25.9	23
1969	98.8	2.5	27.2	25
1970	98.7	2.2	29.2	22
1971	100.7	2.5	31.4	25
1972	104.9	2.4	33.6	23
1973	109.8	3.9	36.6	36
1974	122.8	9.2	41.3	75
1975	137.5	9.7	46.8	71
1976	151.7	11.6	51.2	77
1977	146.7	11.9	55.4	81
1978	150.3	10.8	59.6	72
1979	153.1	14.2	65.3	93
1980	159.6	23.6	72.4	148
1981	169.8	32.7	79.4	193
1982	184.4	34.6	86.0	188
1983	190.4	33.8	90.4	178
1984	195.0	36.8	95.4	189
1985	187.0	36.4	100.0	195
1986	185.0	14.9	103.2	81
1987	173.0	15.6	105.9	90
1988	169.5	12.4	110.4	73
1989	166.3	16.0	116.2	96
1990	163.6	16.9	121.9	103

Source: European Farming Union (1992, unpublished)
[1] Commercial ecu t^{-1} effective support price less co-rep. levy.
[2] Commercial ecu t^{-1} on c.i.f. Rotterdam oil price.
[3] GDP deflator published by Eurostat for EC 12.
[4] Kilograms of wheat per barrel of oil.

illustrative way of looking at the price relativity of the two products. It does not detract from the argument that ARMs are becoming cheaper relative to oil. Part of this is due to the reduction in agricultural prices and the other part to the real increase in oil prices. This trend looks set to continue, since on the one hand oil resources are finite and must eventually rise in price, and on the other agricultural prices are being reduced. At the Uruguay Round of the GATT negotiations, the world's leading agricultural producers agreed to reduce agricultural support prices significantly. The effect of this agreement will be to reduce EC agricultural prices by approximately one third from their 1986 levels.

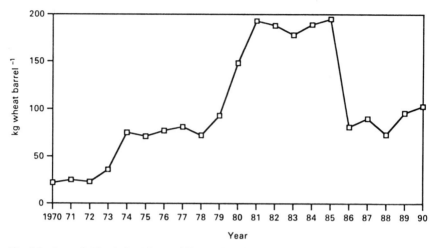

Fig. 4.2 Amount of feed wheat that could be purchased with the value of a barrel of crude oil, 1970–1990 (data from Table 4.6).

THE RANGE OF NON-FOOD USES

There is an enormous range of non-food uses, but a broad distinction can be made between 'speciality' and 'bulk' uses. Specialities include products such as enzymes and antibiotics which command a very high price in the market place and yet require only a small quantity of the agricultural raw material, whereas bulk chemicals are of low value but require a high volume of the agricultural raw material. There is an important trade-off here for agriculture between value added and volume used. The ideal product from the manufacturer's point of view would be one of high value and high volume, but at present this combination does not exist; the nearest equivalent is a product of medium value and medium volume, such as biodegradable plastic. Fig. 4.3 illustrates this price/volume relationship for a variety of products made by the process of fermentation. They range from very high-value products such as diagnostic enzymes and vitamins, which require only a small volume of the agricultural product, down to products of relatively low value such as beer, which require a large volume of the agricultural raw material.

Price/Volume Relationships in Non-food Production

The importance of the cost of the raw material in the overall costs of production varies considerably according to the volume and the value of the product. Table 4.7 illustrates the range of fermentation products

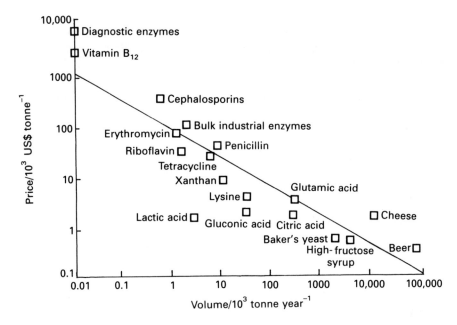

Fig. 4.3 Price/volume relationship for fermentation products (Drozd, 1987).

Table 4.7. Range of volumes and values for fermentation products.

Product	World annual production/t	UK bulk price/£ t^{-1}
Beer	80,000,000	280
Cheese	11,000,000	1,300
Baker's yeast	1,750,000	460
Citric acid	300,000	700
Lysine	25,000	1,900
Penicillin	10,000	45,000
Vitamin B$_{12}$	6	2,300,000

Source: Chemical Industries Federation pers. comm.

by cost per tonne in the UK. Some examples mentioned here are examined in more detail later in the chapter.

Chemicals

Speciality chemicals are typically of low volume and high added value, such as the example of an antibiotic discussed below. They are the much sought-after product of research by chemical and pharmaceutical companies and profoundly affect the profits and share value of those companies; for example, the profits made from the *Herpes* treatment drug Zovirax by the Wellcome Foundation account for half the company's total profits. Hence research emphasis is placed on products such as these. The data in the following sections were prepared by ICI (1986).

Antibiotic

Selling price	£60,000 t^{-1}	
Production cost for 50 t year^{-1}:		
Cost per tonne with sugar at	£100	£350
Variable costs/£	1110	3900
sugar		
other raw materials (RMs)	2230	2230
services etc.	6280	6280
Fixed costs/£	15,000	15,000
Return on capital/£	20,000	20,000
Total costs/£	44,620	47,410
Sugar as % of total	2.5	8.3
Sugar consumption/t year^{-1}	45	45

Yeast

At the other end of the spectrum is a product such as yeast, which is produced in large volume for animal feed. Its price is constrained by the competing feed product, soya, over which it commands a slight premium, but it is still not a product of high value. When the product is of low value it is tempting for the chemical companies concerned to use molasses instead of sugar but this can prove a false economy as the impurities in molasses can cause effluent problems.

Selling price	1.3 × soya		
Current price	£170 t^{-1}		
Trend price	£325 t^{-1}		
Production cost for 10,000 t/year^{-1}:			
Cost of sugar/t^{-1}	£100	£200	£350

Variable costs/£			
sugar	147	294	515
other RMs	159	159	159
Fixed costs/£	83	83	83
Return on capital/£	83	83	83
Total costs/£	472	619	840
Sugar as % of total	31	47	61

Citric Acid

Another product which is positioned somewhere between the two preceding examples but which remains sensitive to the cost of the raw material is citric acid.

Selling price	£990 t^{-1}		
Production cost for 10,000 t year^{-1}:			
Cost of sugar/t^{-1}	£100	£200	£350
Variable costs/£			
sugar	170	340	695
other RMs	134	134	134
services	39	39	39
Fixed costs/£	192	192	192
Return on capital/£	160	160	160
Selling expenses/£	106	106	106
Total costs/£	801	971	1226
Sugar as % of total	21	35	48
Sugar consumption /t year^{-1}	17,000		

Even with the price of sugar as low as £100 t^{-1}, the raw material costs still account for a fifth of the total production cost. For this reason citric acid production has in many cases been relocated to the areas where sugar can be obtained most cheaply.

Fuels

Reduction of production costs through economies of scale are crucially important for fuel uses of ARMs, on account of the relative cheapness of the oil-derived competing products. In order to assess the likely uptake of ARMs for non-food use it is interesting to look in more detail at examples of where substitution of NARMs by ARMs is possible. An obvious example of this is in the energy field. This is already being done on a large scale in Brazil and the USA where cereals and sugar are used to make alcohol (known as 'bioethanol') which can be used either as a petrol additive as in the USA or as the principal fuel as in Brazil. As far as the European Community is concerned, small amounts of fuel alcohol are produced for commercial use in France; sugarbeet and wheat are the principal raw materials used.

THE ECONOMICS OF ARMs FOR ENERGY USE

Bioethanol can be successfully used as an octane booster, and when added to petrol in the proportion of 5% can restore two octane points lost by the removal of lead. If 5% bioethanol were added to petrol in the European Community, there would be a potential market for 11 Mt of cereals and 2.5 Mt of sugar (1 Mt = 10^6 t) if these raw materials were used in a ratio of 2:1. The logic for using a beet/wheat mix is to combine the advantages of sugarbeet, which has a higher yield of ethanol per hectare, with those of wheat which has a high-value protein by-product, vital wheat gluten. Wheat can also be stored for longer periods than beet, thus permitting ethanol plants to function all round the year. A further but less critical reason is provided by the need to maintain a balance between cereal and root crops in the arable rotation.

There is a huge variation in the cost of bioethanol from the various agricultural sources. The cost is primarily dependent upon the yield of ethanol from the various raw materials and, as can be seen in Table 4.8, sugarbeet produces one of the highest yields of ethanol at 53.5 hl ha^{-1}. (The reason why the figure for raw material yield per hectare is the same as that for ethanol yield per hectare is just that it takes one tonne of sugarbeet to produce a hectolitre of ethanol.) Jerusalem artichokes are included in this table because they have been tested for use as a raw material for bioethanol production in France, but in practice they are not widely used since they are difficult to harvest.

The overall cost of producing bioethanol also depends on factors such as transport and processing costs, as well as the value of by-products. From Table 4.9, it can be seen how these factors interact to affect the selection of an optimum crop-mix for bioethanol production.

The oil industry objects to the concept of a subsidy being allowed for bioethanol, and is fundamentally reluctant to become dependent on a subsidized agricultural product for a part of its production process.

Table 4.8. Ethanol yields from various raw materials.

Raw material	Yield/t ha^{-1}	Ethanol yield/hl ha^{-1}
Beet	53.5	53.5
Wheat	5.5	20.0
Maize	6.6	24.0
Potatoes	29.5	31.0
Jerusalem artichoke	66.0	59.0

Source: EEC Commission (1985), *Rapport de synthèse: bioethanol* (unpublished)

Table 4.9. Cost of producing ethanol /ecu hl^{-1}.

	Potato	'B' beet	Wheat	Maize
Raw material cost	45	25	54	68
Transport cost	3	4	–	–
Processing cost	30	20	20	20
By-product value	5	–	20	20
Total	73	49	54	68

Source: EEC Commission (1985), *Rapport de synthèse: bioethanol* (unpublished)

Table 4.10. Bioethanol and its competitors/ecu hl^{-1}.

	Market price	Production cost
Bioethanol	20–35	49–63
Methanol	15–16	13–15
TBA[1]	36–38	29–32
MTBE	42–47	28–33

Source: EEC Commission (1985), *Rapport de synthèse: bioethanol* (unpublished)
[1] Precursor of ETBE (ethyl *t*-butyl ether).

Objections are raised that ethanol costs more to produce than its competitors of fossil fuel origin, and this is certainly true at present (Table 4.10). The most sophisticated competing product produced from oil is MTBE or methyl *t*-butyl ether, the market price of which is above that of bioethanol. However, it is not available in sufficient quantities to meet Community demand for fuel components known as oxygenates, which are designed to raise the octane content of unleaded petrol. Furthermore, methanol, an alcohol derived from oil which is one of the key components in MTBE synthesis, is prone to fluctuations in supply, resulting in sudden price increases. For example, the price of MTBE doubled between the first quarter of 1986 and the first quarter of 1989 from $200 per barrel to $400 per barrel (de Haan, 1989). For this reason, the oil industry has been considering a substitute product ETBE, which is made from ethanol instead of methanol, and this may provide a good opportunity for bioethanol to be used as a co-solvent in this product.

As can be seen from Table 4.10, which shows the price of bioethanol and its competitors, a subsidy is required to make bioethanol attractive in the market place – or, to be more accurate, to enable market entry, given the amount of resistance to the product. French farmers have some practical experience of trying to sell bioethanol as an additive for lead-free petrol and have been successful in persuading the oil companies Total and Elf to use it for that purpose, albeit as a component of ETBE (45% ethanol, 55% isobutene). They have benefited from a French law that allows petrol containing bioethanol to be sold at the same price as diesel fuel, which is 34% cheaper than petrol (3.40F l^{-1} compared with 5.16F l^{-1}). However, this discount still does not cover the whole difference between the cost of production and the market price.

Further assistance was provided by the decision of the French Government in December 1991 to promote biofuels by exempting them from the Internal Tax On Oil Products (TIPP). This tax advantage of 3.23F l^{-1} for a blend with leaded top-grade petrol, or of 2.83F l^{-1} for unleaded top-grade petrol, should enable the bioethanol producers to supply 1–2 million hl of ethanol for fuel use. This measure has been prolonged under the Finance Law for 1993. As a result, Elf has constructed a 450,000 hl year^{-1} ETBE manufacturing plant in Feysin, France. Total is discussing with agricultural producers the construction of three ETBE plants to produce 1 million hl year^{-1}. The French government also called upon the former President of Renault, Mr Levy, to look into the possibilities of developing the use of biofuels. The report was completed at the end of January 1993 but limited its scope to the agricultural problems involved. It suggested that biodiesel was a better solution to the problem of agricultural over-production, simply because it would make use of more land than the production of bioethanol.

At European level, there is at the time of writing a directive under discussion in the European Parliament on the taxing of biofuels, known as the Scrivener Directive after the Commissioner concerned. It proposes that fuels of agricultural origin should benefit from a 90% tax cut. However, the scope of this Directive has been scaled down to apply to the addition of between 0.5 and 1% tax to petrol used in the Community. This Directive is also under discussion in the European

Council where opinion is divided; Britain, Denmark and The Netherlands being opposed to preferential taxation for biofuels.

More recently, the European Commission has undertaken a study of the economics of biomass uses in an interservice working group within the Commission itself. The group worked on producing average Community economic data matrices comparing the costs of production of the biomass fuel (from the cost of the raw material, through transformation, deduction for the value of co-products etc.) with the substitute fossil fuel replaced (adjusted for any differences in energy value and other factors such as an estimated coefficient for the costs of market entry and another factor to discount distribution and profit margins).

It is difficult to undertake this kind of study because:

- agricultural conditions and yields vary in every member state and region of the Community; calculations based on average yields inevitably mask large regional differences;
- the valuation of the co-products is complicated by the absence of markets for some of these; for others it is a question of substituting imported products;
- transformation costs depend on the size of the processing unit, and on the technology the feedstock used;
- the appropriate cost of the raw material in the post-CAP reform world is not easily determined, neither is the world price of most agricultural products stable or predictable;
- the price of the substitute fuel varies across the Community, particularly on account of the variance of excise taxes.

The figures in Table 4.11 should be interpreted with caution. One area of uncertainty is whether the value of the co-products would remain as high if large quantities of ethanol were produced. However, the gap between the value of the substitute energy product and the production cost of ethanol clearly falls substantially as the raw material cost is lowered; indeed, there is an overlap where the raw material is made available at variable cost. Even at world price (Scenario II) the fiscal incentive required is well within the compass of the average weighted Community excise tax on petrol, which is about 43.9 ecu hl^{-1}.

The Commission undertook a similar exercise for biodiesel although the calculation is even more complicated, with conflicting data. Only two scenarios were used: Scenario I, for EC and World market prices which are virtually the same, and Scenario II, an approximation of variable costs. This calculation (Table 4.12) shows the gap between the agricultural and the fossil fuel product to be wider than for ethanol.

The Commission also did this type of calculation for short-rotation

Table 4.11. The economics of ethanol use in the EC/ecu financial.

	Scenarios		
	I	II	III
Wheat[1]			
raw material price	103 t^{-1}	80 t^{-1}	60 t^{-1}
transformation, storage, transport	15–21 hl^{-1}	15–21 hl^{-1}	15–21 hl^{-1}
co-products	10–13 hl^{-1}	10–13 hl^{-1}	10–13 hl^{-1}
total production cost	40–31 hl^{-1}	33–24 hl^{-1}	28–19 hl^{-1}
Sugarbeet[2]			
raw material price	27.7 t^{-1}	14 t^{-1}	10 t^{-1}
transformation, storage and transport costs	10–17 hl^{-1}	10–17 hl^{-1}	10–17 hl^{-1}
co-products	1.5 hl^{-1}	1.5 hl^{-1}	1.5 hl^{-1}
total production cost	43–36 hl^{-1}	30–23 hl^{-1}	25–19 hl^{-1}
Approximate value of substitute energy product:			
Mogas FOB Rotterdam	17.5 hl^{-1}	17.5 hl^{-1}	17.5 hl^{-1}
Average EC Mogas (ex tax)	23.8 hl^{-1}	23.8 hl^{-1}	23.8 hl^{-1}
Average EC excise tax	43.9 hl^{-1}	43.9 hl^{-1}	43.9 hl^{-1}
Substitute energy value	19–13 hl^{-1}	19–13 hl^{-1}	19–13 hl^{-1}

Scenario I   EC market price reflecting CAP reform decisions
Scenario II   Approximate world market prices
Scenario III   Variable costs

[1]Conversion factor 1 t wheat = 3.6 hl ethanol.
[2]Conversion factor 1 t beet = 1 hl ethanol.

Table 4.12. The economics of EC biodiesel (diester) production/ecu financial.

	Senario I	Scenario II
Price of rape grains	144 t^{-1}	120 t^{-1}
Transformation, storage	172 t^{-1}	172 t^{-1}
Glycerine co-product	10–30 t^{-1}	10–30 t^{-1}
Oilcake	150 t^{-1}	150 t^{-1}
Total production cost	35–33 hl^{-1}	35–33 hl^{-1}
Spot price Rotterdam diesel	15.3 hl^{-1}	15.3 hl^{-1}
Average EC selling price	20.7 hl^{-1}	20.7 hl^{-1}
Average EC excise tax	28.5 hl^{-1}	28.5 hl^{-1}
Approximate value of substitute energy	14 hl^{-1}	14 hl^{-1}

Conversion factor: 2.55 t rape = 1 t diester;
                  1 t diester × 0.8 = 1 hl diester.

Table 4.13. The economics of wood or lignocellulose as substitutes for domestic or tertiary heating/ecu financial.

	Senario I	Scenario II
EC market price/ecu t^{-1}	21	12
Transformation, storage, transport	23	15
Co-products	0	0
Total production costs	44	27
Equivalent fuel/ecu/toe* (0.43 t^{-1} wood)	102	63
Equivalent value of domestic fuel	8.7	5.4
Approximate value of substitute product	17.85	17.85
After distribution (0.5) margins, efficiency and competition (0.8)	7.14	7.14

*toe = tonnes of oil equivalent

forestry or lignocellulosic C_4 plants as a substitute source of energy for domestic and tertiary heating (Table 4.13). Scenario I refers to the present price of the agricultural raw material and Scenario II the price in the medium term.

In these two scenarios the gap is already small under Scenario I, but negative in the medium term under Scenario II. For short-rotation forestry, the long-term outlook is particularly attractive providing the technical problems can be overcome to increase efficiency of gasifying the woody residues. Sorghum, *Miscanthus* and other products could also turn out to be attractive options in the medium term.

The experience of biofuel producers elsewhere in the world is not fundamentally different. In the USA there is also resistance to bioethanol from the large oil refiners. However, because lead-free petrol has been available for longer in the USA than in Europe, and also because of the political power of the large maize-producing states, a large quantity of bioethanol is actually sold and blended in America.

The case for using bioethanol was assisted when the US Congress passed a Clean Air Act in November 1990. This formally recognized the need to reduce the level of pollution for fuel emissions by using environmentally benign fuel substitutes. The removal of lead and other toxic materials such as the aromatics (benzene, toluene and xylene) greatly increases the demand for fuel oxygenates, of which ethanol is one. Recognizing the risk of a shortage of oxygenates, the US government set up an Oxygenated Fuels Programme which came into effect in 1992. The preparations for this consist of many discussions between the fuel refiners and motor car manufacturers to see what can be done to ensure an adequate fuel oxygenates supply. The Environmental Protection Agency (EPA) considers that there will probably be just enough fuel oxygenate available but that the supply situation is tight; they estimate that a quarter of oxygenate demand will be met by ethanol from maize, but that this amount may dwindle as supplies of the more sophisticated product MTBE come on stream. The petroleum refiners and the motor manufacturers consider that there are sufficient supplies of oxygenates available without having to pay US$1.10–1.20 per gallon of ethanol which is a hydrocarbon of the olefin series. The fact that this still appears a viable market for bioethanol production is demonstrated by the recent decision of the US grain trader, Cargill, to construct ethanol production plants. However, these will still be on a small scale of under 30 million gallons a year in order to take advantage of a tax credit designed to help the small producers.

In the USA, the large corn-producing states have succeeded in obtaining a tax waiver for ethanol blended with petroleum. This tax waiver appeared to be under threat as the Bush administration sought to reduce its budget deficit, but after a year of negotiations the ethanol producers have succeeded in getting the Federal tax exemption extended albeit at a slightly lower rate. Henceforth, the tax exemption will fall from 6 cents per gallon for the blender to 5.4 cents.

Overall US agricultural policy is not expected to be changed fundamentally under the presidential leadership of Bill Clinton, given his pre-election promises to keep the status quo. Various groups are urging him to support projects for biofuels, such as an engine optimization programme for biofuels to increase the mileage of cars and light pick-up trucks to 45 miles per gallon (about 16 km l^{-1}) equivalent. This development is already under way, involving the EPA and the Department of Energy. With additional funding, these engines could be available commercially by 1995. Other programmes that need support include research aimed at reducing the costs of production of biodiesel and ethanol. Support is also needed for fuel performance and fuel characteristics research to gain the approval of the EPA. At the time of writing, the Clinton administration has still to decide on its research priorities.

In Brazil the situation is somewhat different, because the Brazilian government took a decision some time ago to encourage the use of 100% ethanol in vehicles by differential taxation. This policy was so successful that most cars in Brazil are now run on ethanol produced domestically from sugar cane. In fact Brazilian sugar cane plantations meet demand from two sources: the world sugar market and the fuel alcohol market. This created problems in 1989 when there was a boom in world sugar prices at the same time as an increase in fuel alcohol consumption. Originally the new Brazilian president had pledged to reduce the dependency on fuel alcohol, but the 1990 Gulf crisis led to a U-turn in policy. By the end of August 1990, he had decided instead to expand the fuel alcohol programme and announced three initiatives:

- a US$300M credit to the cane industry;
- the adjustment of fuel alcohol prices in line with inflation;
- the creation of a commission to study the debt problem of the cane sugar alcohol producers.

The first of these initiatives has already been acted upon. Cane alcohol prices have been increased, but still lag behind inflation by 5–6 months. There are about 400 cane alcohol producers in the northeast of Brazil with long-standing debts – dating back to the 1950s in some cases – but the debt initiative will be directed at the newer, more efficient producers.

Since these initiatives were decided upon, production of alcohol has proved to be 0.5 Mt higher than expected and consumption 0.5 Mt lower. The government has therefore created a 1 Mt strategic ethanol reserve.

MACROECONOMIC ADVANTAGES OF USING ARMs

Environmental Advantages

The relative pricing of oil-derived and agriculturally derived oxygenates suffers from the same problem as that of their raw materials, the failure to take account of their full social costs and benefits. There is one fundamental difference between fossil energy and energy derived from ARMs, and that is the latter's renewability. There are several significant environmental advantages of using ARMs, of which the most obvious is the impact on the so-called 'greenhouse effect'. This is the result of the rise in the amount of carbon dioxide (CO_2) in the Earth's atmosphere, which may cause an increase in temperature that in the long run could have disastrous consequences for the planet's environmental equilibrium. Carbon dioxide is released in large quantities when the hydrocarbons laid down under the Earth's crust over millions of years are extracted and burned within a very short space of time (CEA, 1988).

By contrast, the CO_2 liberated when ARMs are burned is re-absorbed by replacement plants in the process of photosynthesis (except, of course, where large areas of biomass are burnt and not replaced). Thus there is no net increase of CO_2 in the atmosphere. This natural carbon cycle provides a fundamental scientific justification for promoting the use of ARMs over hydrocarbons. Furthermore, the planting of short-rotation forestry units for biomass production could actually help to counteract the rise in CO_2 levels, because the intensive production of biomass causes a higher volume of CO_2 to be stored.

Using ARMs as an energy source brings other environmental advantages compared with using fossil sources. Sulfur dioxide formed during the combustion of oil is held by some to be one of the principal agents of the damage caused by 'acid rain'. Since there is virtually no sulfur in biomass, there is little risk of producing sulfur dioxide when biomass is burned. The amount of nitrogen oxides (another cause of 'acid rain') produced when ARMs are burned is also lower than for hydrocarbon fuels. When biomass is burned some aldehydes are produced, but not the toxic formaldehyde as in the case of hydrocarbons. Finally, since bioethanol made from ARMs enables lead to be removed from petrol it offers another significant environmental advantage.

Security of Supply

Other considerations may also influence a decision in favour of ARMs for energy use. One of the most important is the strategic argument of security of supply. Apart from the UK, no member state of the European Community has substantial domestic oil reserves. Most of the world's oil resources are concentrated in the Middle East, an area not noted for political stability. The experience of the two oil price shocks of the middle and late 1970s was damaging for the economies of western Europe. However, since the subsequent reduction in oil prices the EC has not considered it necessary to take drastic measures for the promotion of alternative energy sources. This decision has come under review since the Gulf crisis of 1990, which demonstrated how suddenly and unexpectedly political events in the Middle East can endanger supplies of oil to the developed world.

Balance of Trade

A directive of the European Council of 5 December 1985 concerned the possible economies which could be achieved by substituting imports of crude oil, but it has only led to demonstration-scale projects for energy from wind, water and the Sun. Yet the addition of 5% of bioethanol to petrol could spare the Community imports of 2 Mt of crude oil. At current oil prices that is the equivalent of 543M ecu. There could be further advantages for the Community balance of trade, because the by-product of bioethanol produced in this quantity could save the Community 4 Mt of protein feed imports worth 857M ecu. Of course, the balance of payments effect is not an improvement of 1.4 bn ecu, for the saving would be offset by additional imports of agricultural inputs and the relative efficiency of agriculture versus other sectors should be considered. Nonetheless, there is some apparent first-round balance of payments advantage.

Employment

Beyond the balance of trade effects of bioethanol production, there are other macroeconomic effects, notably upon employment. One of the dilemmas facing European governments is the sensitivity of the electorate to unemployment. Political decisions to take agricultural land out of production would not be without consequences for jobs in areas where employment alternatives are scarce. By contrast, bioethanol production would not only maintain job opportunities, but actually create new ones. The reason for this would be the need to construct bioethanol production units in rural areas, on account of the transport costs of the ARMs. Ancillary industries would also benefit economically from this additional economic activity in rural areas, whereas the reverse is true where land is taken out of production.

It has been estimated that for the production of 1.5–2.5 Mt of ethanol between 23,600 and 39,300 jobs would be created (Agro-developpement *et al.*, 1987). Set against this, there would be some loss of employment in the oil sector due to the displacement of a proportion of oil products by bioethanol.

Fiscal Benefits

One of the strongest arguments for providing a tax incentive biofuel is the overall fiscal benefit which accrues from bioethanol production. The production of biofuels increases investment and economic activity. This in turn generates tax income and contributions to social organizations. From a budgetary perspective, the replacement of current unleaded petrol by petrols of equivalent quality containing biofuels enables budgetary savings to be made. Without generating greater activity, employment and investment, current unleaded petrol costs the budget three times more than petrol with a bioethanol or ETBE additive. The difference arises because the tax benefit afforded to unleaded petrol is granted on the total volume of this petrol, whereas in the case of biofuels (these being used as additives), the tax benefit only involves the biofuel fraction of the petrol.

International Trade Relations

It is not just domestic considerations which will influence the decision-making on the use of ARMs, but also external factors. There is increasing pressure from the Community's trading partners, notably the USA, for European agricultural policy to be adapted to allow greater scope for free trade. This was one of the prominent issues of the Uruguay Round of negotiations on the General Agreement on Tariffs and Trade (GATT). These negotiations were aimed at reducing the level of agricultural support worldwide. Although agricultural energy was not specifically on the GATT agenda, a reduction in the price of ARMs could help the competitiveness of ARMs for energy use. Furthermore, the emphasis on free trade might encourage access to markets for agricultural alcohol.

The Economics of Other Non-food Uses

ARMs can substitute for oil not only for energy purposes, but in a whole range of other industrial applications too, especially in the chemical industry. ARMs can substitute for oil to provide the basic building blocks for the synthesis of chemicals such as organic acids and alcohols, and they can be used in the manufacture of speciality products such as enzymes and antibiotics. The economics of this wide range of products varies enormously, from speciality chemicals for which the cost of the raw material is insignificant compared to the value of the end-product down to bulk chemicals which are extremely price-sensitive to the cost of the raw material. This is well illustrated by the example of biodegradable plastic.

Already, commercial production has begun of a biodegradable plastic known as polyhydroxybutyrate (PHB) or 'Biopol'. ZENECA (formerly ICI) have developed a process whereby a microorganism uses a carbohydrate substrate to produce a fat from which plastic can be made. PHB is capable of substituting existing polypropylene plastic (made from oil), and yet has the additional property of biodegradability. At present, 1 Mt of oil-derived, non-degradable plastic are consumed annually in the European Community. At least some of this market could be captured for PHB provided that the carbohydrate substrate is supplied at a competitive price level. Conventional plastics derived from oil range in price from 2500 to 3000 ecu t^{-1}. In order to compete with the lower price of oil-derived plastics, the carbohydrate substrate would need to be supplied at no more than 320 ecu t^{-1}, or 60% of the intervention price of white sugar. At present, the product is being manufactured in small quantities for high added-value purposes, such as medical applications. However, a much larger-scale use may emerge if other European governments follow the Italian example of making biodegradable plastic mandatory for use in wrapping. This measure was given even greater force by an Italian law effective from 1 February 1989, imposing a tax of 100 lire on all non-degradable plastic bags. This is a clear example of the determination of policy by considerations other than purely commercial ones, which could have a substantial impact on the uptake of ARMs.

Outlets for Sugar and Starch

The CAP imposed a system of production refunds on the chemical industry governing their purchase of EC agricultural raw materials. However, in most years the system tended to make EC raw materials uncompetitive compared with those available in third world countries. In 1980, the European Council of Chemical Manufacturers' Federations (CEFIC) began making strong representations that they should be allowed to purchase their raw materials from the Community at more competitive prices. At that time, the regimes covering the sale of sugar and starch to the chemical industry allowed the end-user to receive a production refund, but in the case of sugar this only covered 10% of the difference between the world market and Community prices. Not surprisingly, the consumption of sugar for industrial non-food uses actually declined during the early 1980s (Table 4.14).

On the eve of the reform of the sugar and starch regimes which govern sales to the chemical industry, the volume of sugar for which production refunds were sought had dwindled to just over 60,000 t. The discrepancy in the system of production refunds led to a ridiculous situation, where EC chemical companies were transferring their production capacity outside the Community in order to benefit from the full export refund on Community agricultural surpluses. Given that the sugar export and production refunds for chemical uses are financed largely by production levies, EC beet growers were actually paying to deprive themselves of a market; indeed, because of the way the EC sugar regime operates one could argue that consumers have been paying for this absurdity.

The optimistic forecasts by the chemical industry of the uptake of agricultural raw materials encouraged the farm lobby to lend its weight

Table 4.14. The consumption of sugar by the European chemical industry/tonnes for which a production refund was applied.

	1978–79	1984–85
Organic acids	11,057	8,709
Organic alcohols	12,741	8,994
Fructose	5,670	10,861
Penicillin	25,804	14,885
Other antibiotics	5,591	665
Adhesives	1,023	695
Other conventional chemicals	20,386	17,283
Total	82,272	62,092

Source: CEFIC (1985)

Table 4.15. Projected consumption of sugar by the European chemical industry (metric tonnes).

Organic acids	168,000
Organic alcohols	–
Fructose	14,000
Penicillin	40,000
Other antibiotics	8,000
Adhesives	4,000
Other conventional chemicals	100,000
New biodegradables	200,000
	534,000

Source: CEFIC (1985)

to the chemical industry's efforts to reform the regulations. These were successful and in 1986 it was decided to increase the refunds for certain chemical products to make EC raw materials more competitive. The end-products eligible for refunds qualified only if they enjoyed no other form of protection under the CAP. Thus, for example, food products such as glucose and levulose were removed from the list whilst some new products such as biodegradable plastic were added to it.

The chemical industry anticipates the strongest growth to come from a new generation of biodegradable products. Their detailed estimates of sugar consumption by the mid-1990s (Table 4.15) show how important this category of products could become.

There has been an increase in annual sugar consumption by the EC chemical industry from around 60,000 t before the reform of the regime to 189,000 t in 1991–92. (In order to compare like with like, the figure of 10,000 t for Spain's consumption of sugar prior to the reform in 1986 should be included in the prereform sugar consumption levels, taking them to approximately 72,000 t.) Since the reform consumption has dropped back a little on account of the increase in world sugar prices, which make sugar less attractive to use than glucose.

ALTERNATIVE FORMS OF SUPPORT FOR NON-FOOD USE

It should be clear by now that for many of these non-food uses to be implemented, some kind of support or subsidy may be necessary in order to get production started. Different methods are used around the world to achieve this. One of the most common is subsidy or fiscal exemption. This is practised in the USA to encourage the production of bioethanol. In Europe, a variety of measures have been tried over the years. One of the obstacles has been the two-tier pricing system for agricultural products, in which the domestic agricultural prices are above world prices. For non-food uses to begin to be competitive it is necessary to subsidize or refund these prices down to world market level. This can be done either by passing legislation to allow raw materials to be sold at world market price for non-food use, keeping them completely separate from the supply of agricultural raw materials for food use, or by an elaborate system of refunds for specific products as practised in the European Community.

Production Refunds

Until 1986, the EC operated a limited system of production refunds for certain listed products, the raw materials of which benefited from a restitution that in theory took their price down to world market price. However, due to shortcomings in the method of calculating the refunds, they frequently only covered 10% of the difference between Community and world market prices. Naturally this acted as a disincentive to the development of non-food uses in Europe. A major reform of the regime in 1986 (discussed in more detail above) has allowed a significant increase in the volume of ARMs used for non-food purposes. A further restriction is caused by the limited number of products upon the list – a so-called 'positive list' – which are eligible for refunds; for example, bioethanol is excluded. A further weakness of the refund system lies in the reluctance of some industrial manufacturers to give away confidential information about their potential products by declaring them on an official list. It is also extremely bureaucratic as proof has to be given that the refund was only used for a listed product. The system is therefore inherently cumbersome and not conducive to the extension of non-food uses.

Industrial Set-aside

A new opportunity to promote non-food uses has presented itself in Europe with the extension of the set-aside scheme, which allows a farmer to be paid compensation for taking land out of food production. The original proposal for so-called 'industrial set-aside' allowed a farmer to grow certain crops, notably oilseeds and cereals, for industrial purposes on land taken out of food production. In return he or she would receive a reduced rate of compensation, because obviously the non-food crop grown would have some commercial value. The original proposals were made by the European Commission in 1991 and amended and adopted by the Council of Agricultural

Ministers in May 1992. The amendments required farmers to set aside 15% of their arable land. This land set aside would then be compensated at the rate of 55 ecu t^{-1} multiplied by the theoretical cereal yield of the region. These changes also allowed for oil crops to be grown for industrial purposes.

The scheme has still not proved particularly popular with farmers and so the European Council decided in June 1992 to allow all crops grown for non-food use to receive the full set-aside payment, which is equivalent to £208 per hectare. The crops which can be grown for industrial purposes include short-rotation forest trees, potatoes, peas, rye, barley, oats, maize, sorghum, wheat and other cereals except those grown for seed, rape, linseed, groundnuts, soya, cotton, castor oil seeds, sunflower, flax, and a range of plants for perfumery, insecticidal, fungicidal and plaiting, stuffing and padding purposes. Most recently, sugarbeet has been added to the list following a Council meeting in May 1993.

Crops grown on set-aside land will only be eligible for payment if their primary end-use is one specified within the regulation. In general permitted non-food uses include the production of crops for processing into industrial oils and fats, the production of ethanol for use in motor fuels, and the production of crops for burning in power stations. The growing of crops for pharmaceuticals is also permitted. In addition for a specified non-food use, the non-food products obtained from the processing of the raw material must be greater than the value of all other by-products of that process which may be used for human and/or animal consumption. In order to obtain the compensation available under the set-aside scheme, the claimant must be able to prove he or she has a contract concluded between the producer and the first processor specifying the principal end-use of the raw material. Claimants also have to submit a sowing declaration that shows the species and variety of the crop grown for non-food purposes. National governments within the EC retain the power to prohibit the use or production of any crop grown for non-food purposes on agronomic or environmental grounds (National Farmers Union, 1993, internal briefing note).

The scheme remains complicated for farmers to operate and seems a curiously convoluted way of encouraging diversification in agricultural production. As regards fuel uses, it might be better to convert the set-aside premium to an amount per hectolitre of biofuel payable to the fuel producer, and allow the market to decide which crops can be grown most successfully for biofuel production.

Tax Abatements

The most likely tool for achieving the widespread introduction of fuel alcohol in petrol would be using the existing system of fuel taxation. This is already done in the USA and Brazil but not so far in the European Community except in France. The French government has decided to reduce the taxation for petrol containing bioethanol to the same level as diesel fuel, i.e. by 34%; this was the level considered necessary to allow bioethanol to enter the market. Other countries in Europe have so far not followed this model. It is well known that fuel taxation is an accepted method of raising government revenue. In Europe as much as two thirds of the fuel price consists of taxes; therefore governments have a lot of room for manoeuvre. They are, however, reluctant to give tax credits of the order of one third. There is no fiscal harmonization in the European Community as yet, and tax differentials between fuels vary considerably. However, there is a proposal on the table from the European Commission to allow a 90% cut in the tax applied to fuels of agricultural origin. Perversely, the British government is considering a special fuel tax for biodiesel. This just illustrates the need for fuel tax harmonization in Europe. This lack of agreement about the tax status of biofuels works against their commercialization.

Conclusion

The economics of non-food use are complex and inextricably linked with the vagaries of world market prices for key commodities. For many non-food applications, certainly those on a large scale, agricultural raw materials will be in competition with oil-derived products. Thus the development of world oil prices is crucially important for the commercial viability of using agricultural raw materials in place of oil. Oil and agricultural prices would then be linked. However, shocks in the energy market may well have more impact on agricultural prices than that of shocks in the food market on energy prices. Thus if agricultural raw materials were to be increasingly used for energy production their price would eventually impact on oil; that, however, is not likely to be for some time to come. All these commodities are affected by political events, whether political upheavals such as tensions in the Gulf or trade battles such as those which took place over the Uruguay Round of the GATT talks. The underlying economic trend is, however, in agriculture's favour for oil is a finite fossil fuel and will thus eventually rise in price whilst agricultural raw materials are renewable and becoming progressively cheaper relative to oil.

The main difficulty faced by producers of agriculturally derived substitutes for oil-based products is that the latter are almost invariably cheaper at present and look to remain so for the foreseeable future. This implies that subsidies need to be paid if the agriculturally derived product is to enter the market. Naturally, there is resistance to such a political instrument, which involves public money and may fall victim of a change of policy at any time. It has not been possible as yet to persuade the consumer that the macroeconomic and environmental benefits of an agriculturally derived product warrant a premium being paid over and above the price of its oil-derived equivalent. To encourage usage it has been necessary for governments to intervene, as in the case of the tax waiver for bioethanol fuel blends in the USA or the tax system encouraging the use of biodegradable plastic wrappings in Italy.

The breakthrough will probably only occur once the price of oil reflects its true environmental and social cost, and as long as it is so politically influenced by OPEC members this remains unlikely. Thus the short- and medium-term prospects for non-food uses are not very encouraging; however, in the long run the underlying economic trend of the relative cheapening of agricultural raw materials must work in their favour.

REFERENCES

Agro-developpement, Laurence Gould Consultants, Parpinelli Tecnon and Institut fur landwirtschaftliche Energie, University of Braunschweig (1987) *Study on Bioethanol*. Report prepared for the EC Commission.

British Petroleum (1992) *Statistical Review of World Energy*. Britannic Tower, Moor Lane, London EC2Y 9BU, UK.

CEA (Confederation Européene de L'Agriculture) (1988) *Biomasse et Approvisionement Energetique*. CEA Brugg, Switzerland.

CEFIC (European Council of Chemical Manufacturers Federation) (1985) *The Use of Agricultural Raw Materials in the European Chemical Industry*. CEFIC, Brussels.

Drozd, J. (1987) *Hydrocarbons as Feedstocks for Biotechnology*. Fermentation and Microbiology Division of Shell Research (UK), Sittingbourne, Kent, UK.

European Commission (1989) Oil price forecasts prepared by the Directorate General XVII on 5.7.89, Brussels.

de Haan, H.E. (1989) European market developments in oxygenated fuels. Paper read at the 1989 Alcohol Week Conference, Holland.

ICI (1986) Slides produced to illustrate a paper given by D. Stringer at the biennial congress of the International Confederation of European Beetgrowers in Athens, May 1986. Copies available from 29 rue du General Foy, Paris 75008.

International Monetary Fund (1992) *International Financial Statistics Yearbook.* Washington, USA.
Oil and Gas Journal (1992) *Database Survey of Forecasters.* Penwell Publications, Oklahoma.
World Bank (1991) *Commodity Trade and Price Trends.* Johns Hopkins University Press, Baltimore.

5

POLICIES AFFECTING NON-FOOD USES

There are several different policy areas that affect the development of non-food uses of agricultural raw materials. The most obvious of these is the type of agricultural policy which varies in its policy instruments from one country to another. There are remarkable similarities between the agricultural policies of the major agricultural producing nations. They are nearly all 'mature' in the sense that they have achieved the objectives for which they were set up in the first place. They nearly all suffer from the same problem of trying to reconcile social and economic goals for agriculture. There are often contradictions between these twin goals, and the result is a market-distorting policy. These problems will be illustrated in this chapter by taking examples from the European Community, Japan and the USA.

There are other policies such as financing and taxation, those affecting basic and applied research, intellectual property law and training which have some bearing on the progress of non-food uses. The United States Office of Technology Assessment identified several policy areas that affect the future competitive position of different countries with respect to the commercialization of biotechnology. Its list provides a useful tool to help determine the international competitiveness of different countries with respect to the commercialization of non-food uses.

AGRICULTURAL POLICY

Most agricultural policies focus on the need to produce food. Non-food production in agriculture was not considered a high priority in the aftermath of the Second World War, which had resulted in food shortages even in the world's major agricultural producing countries. The ability even to consider using agricultural land for non-food purposes is a result of remarkable growth in agricultural productivity, notably fuelled by improvements in yield. Even former importing countries such as those in Europe are now exporting their food surplus and world markets for agricultural produce have become depressed as a result.

This imbalance in agricultural trade is what prompted the world's major agricultural producers to seek an agreement on the liberalization of agricultural trade. This was the objective of the Uruguay Round of negotiations on the General Agreement on Tariffs and Trade (GATT). After years of negotiation a deal was reached on the deadline date of 15 December 1993. The deal was based on a compromise between the main negotiating parties of the European Community, the USA and Japan, known as the Blair House Agreement, the main features of which are as follows:

- A reduction of tariff equivalents of 36% on average by 1999; however, for sugar, wine, olive oil, skimmed milk powder and certain fruit and vegetables the tariff reduction will be 20%.
- Minimum Import Access increased to 5% of domestic consumption at its 1986–1988 level.
- A reduction in subsidised export volume by 21% of 86–88 base.
- A reduction in export subsidies by 36% of 86–88 base.

108 countries are signatories of the GATT accord; these include the large agricultural producers such as Australia, Argentina and Brazil as well as many smaller and less-developed agricultural nations. So one might expect to see liberalization of agricultural trade and a reduction in the price of key agricultural commodities. A reduction in the level of agricultural support and the reconciliation of support prices with world market prices ought to help the progress of non-food uses by making agricultural raw materials cheaper to use. Easier access to agricultural markets for imports should also help to improve the competitiveness of agricultural raw materials for non-food use. For example, highly protected markets, such as those for the by-products of starch manufacture, ought to become more accessible to cheaper imports. However, it is difficult to forecast the full implications of the liberalization of trade. Paradoxically, a reduction in the by-product credits of starch processing might force an increase in the price of the principal product as manufacturers try to maintain their own profits. No sooner does some progress appear to have been made than the old dichotomy between social and economic objectives in agricultural policy surfaces once again. An illustration of this is the number of compensation schemes in the main agricultural producing countries which will mitigate the effects of the GATT deal. In the European Community, the price reductions will be mitigated by Area Compensation Payments at a fixed rate per hectare depending on crop and region

and subject to a portion of a farmer's land currently devoted to supported crops being left idle. This concept of taking land out of agricultural production is seen as the second major tool to help re-balance world agricultural markets. The GATT deal may have the effect to reduce both the **price** and the **volume** of agricultural production. How effective these arrangements will be is open to conjecture.

Compensation for land set aside will be paid at the rate of £311 per ha. In practice, set-aside will only apply to farmers with a production potential of more than 92 t of cereals. This will apply to 54% of farms in the UK, where farm structures are significantly larger than in other parts of the Community. By contrast, it is estimated that at least 95% of cereal farmers in Greece, Italy, Spain and Portugal will be exempt from the set-aside requirement and that in Belgium, Germany, Luxembourg and Holland, 80% of farmers will be exempt. Even in France and Denmark, only 30% of farmers will be required to set land aside. These arrangements therefore provide no real incentive for Europe's small farmers to reduce output. It is important to note that it is the small farmers who have contributed significantly to the increase in cereal production. In 1991, total EC cereal output increased by 6% to 167.5 Mt. Greece accounted for 35% of this increase, Portugal 26%, Italy and Ireland 10% each and France and Germany 8% and 7% respectively. Thus, it is unlikely that the set-aside restrictions will be particularly effective in controlling the volume of production.

The next section will examine the individual agricultural policies of the three main competitors in the area of non-food uses – EC, the United States and Japan – to see how these policies help or hinder the growth of non-food uses.

European Community

The political factors influencing non-food uses predate the establishment of a Common Agricultural Policy (CAP) in EC, but widespread interest in the development of this sector as a major new outlet for agricultural raw materials is inextricably linked with the CAP and attempts to reform it.

The origins of the CAP have to be understood in the context of post-war food rationing in Europe. Its objectives are set out in Article 39 of the Treaty of Rome of 1957 and may be summarized as follows:

- to increase agricultural productivity;
- to ensure a fair standard of living for farmers;
- to stabilize agricultural markets;
- to assure availability of supplies;
- to ensure reasonable food prices.

There is no specific mention of non-food uses here, but that is understandable when the first priority was to produce enough food at an affordable price. More than 30 years on, it may be said that the CAP has successfully fulfilled these objectives and pressure has grown for a new set of priorities to be established.

In 1985, the European Commission introduced a Green Paper (EC Commission, 1985) which presented a series of political options to enable the portion of the Community budget spent on agriculture to be reduced. Structural surpluses of food had arisen in many sectors, resulting in the need to dispose of them (usually at high cost) on world markets. These are known as 'structural surpluses' because they are built into the production targets themselves. The Green Paper set out a number of ways of balancing agricultural markets. The preferred option was price policy over further quantitative restrictions. The reasons for this were given as follows:

- the fear that more quotas would restrict productive potential;
- the risk of encouraging substitutes with artificially high prices;
- the risk of resistance from consumers;
- the threat to the unity of agricultural markets.

The Commission summed up its recommendation as follows:

there can be no alternative to pursuing a price policy more adapted to the realities of the internal and external markets but taking account of the Community's obligations to the agricultural population.

The second half of that statement indicates some degree of compromise to lessen the severe effects of price pressure if applied without regard for one of the original objectives of the CAP, that of guaranteeing farmers a reasonable standard of living. Undoubtedly, the opportunities perceived for farmers in non-food uses were part of the positive perspective the Commission wished to provide. Whilst this option could not constitute a 'miracle solution' it was hoped that non-food uses would contribute to the re-balancing of agricultural markets.

In May 1988, the Commission reviewed what had been achieved so far by agricultural reform (EC Commission, 1988). By that time, price reductions had been in operation for cereals, oilseeds and beef. Contrary to its original intentions, the Commission had opted for a quantitative restriction to control dairy production because of the extent of the market imbalance and lack of alternative outlets for that sector. There was no specific mention in this document of measures to encourage non-food uses. However, there was a set of accompanying measures designed to mitigate the severity of price pressure in order 'to make it bearable for the most disadvantaged farmers'. These measures were:

- aid for small cereal growers;
- an increase in the compensatory allowances paid to farmers in hilly and less-favoured areas;
- an aid scheme to encourage farmers to stop farming;
- a scheme to provide aid for the withdrawal of land from production (set-aside).

The Commission was forced to reflect on the question of whether its reform went far enough. Its conclusions were that progress could not be too rapid because of the trend in agriculture for the number of farms to decline by 2% per annum. Excessive price pressure or quantitative control would, it feared, come up against insurmountable political and social obstacles. The European Community is starkly different from its major competitors in having a large proportion of its active population still employed in agriculture: 8.3%, as against 3.1% in the USA. In consequence there were few profound changes to the CAP over the five-year period 1985–1990. With the possible exception of budgetary stabilizers (discussed below), there was disappointingly little to foster the growth of non-food uses.

Critics of the Common Agricultural Policy stepped up their pressure for fundamental reform in the face of this mild approach by the Commission. Their main concerns were the danger of the open-ended funding of agricultural policy and the embarrassing surplus of food to which this had led. In response to this pressure the Commission introduced a package of so-called 'budgetary stabilizers' in February 1988. As the title implies, their objective was to stabilize the flow of money from the Community budget towards agriculture. This was to be achieved by imposing a quantitative threshold, known as a Minimum Guaranteed Quantity (MGQ), on certain types of production apt to be in surplus – such as cereals – and then introducing an automatic price cut if the MGQ was exceeded. So, for example, the threshold for cereals was set at 160 Mt for the EC-12 which, if exceeded, would have triggered an automatic price cut of 3%. This was of no immediate help towards the fostering of non-food production as any cereals destined for non-food production had to count towards the MGQ, however, over a period of years, given rising productivity, reduced the real cost of agricultural raw materials.

A declaration of intent to encourage non-food uses was adopted at the same time as the stabilizer package in February 1988. On this occasion, the European Council of agricultural ministers approved a statement which called on the Commission to 'investigate all the possibilities of increasing the utilization of agricultural commodities in the non-food sector and to submit proposals to this effect'.

The political measure chosen by the Commission seems to be tangential to the objectives. The Commission simply proposed to amend the set-aside scheme for taking land out of food production to allow growers to use the compensation for crops grown for so-called 'industrial purposes'. The measure has not been successful in encouraging the wholesale diversification of sections of arable farms into non-food production because:

- the level of compensation is too low;
- it is complicated for the farmer to apply;
- investors fear the measure could easily be withdrawn.

Another major attempt to reform the CAP had some success in June 1992. Under the twin pressures of a desire to reduce agricultural expenditure within the European Community and pressure from Europe's trade partners to force a reduction in the level of agricultural support, the European Council of agricultural ministers reached a decision to cut the institutional price levels of most of the key arable commodities. This decision was accompanied by a scheme to compensate farmers for loss of income providing 15% of their arable land is taken out of food production. There are two principal effects on the prospects for non-food uses. The first is indirect, through the lowering of the institutional prices, making the agricultural raw materials cheaper; the second is through the mandatory set-aside measure, which would enable crops to be grown for non-food purposes on land retired from food production.

In parallel with the interest in non-food uses shown by the Agricultural Directorate (DG VI) of the European Commission, there were other political developments that encouraged non-food uses in the Directorate for Science, Research and Education (DG XII). Two reports were published by special teams set up within DG XII (EC Commission, 1986 and 1987). These culminated in a project known as the ECLAIR programme, the name of which is an acronym for European Collaborative Linkage of Agriculture and Industry through Research. The project is aimed at improving the interface between agriculture and industry, its overall objective being to remove the scientific obstacles to the use of EC agricultural resources by industry and in so doing to forge a lasting partnership between the parties. One strong motivation for this was the fact that Europe lagged behind the USA and Japan in these developments. It co-finances applied (but still precompetitive) research in the area of agroindustry, including all kinds of primary production of biological resources and the related up- and down-stream industrial processes. The call for proposals was launched on 17 December 1988; 220 proposals were received, requesting funding of up to 400 million ecu from the Community.

The total budget for ECLAIR is only 80 million ecu of which 15 million ecu are earmarked for the management of the programme, workshops, studies and training grants. A selection of 42 of the projects therefore had to be chosen. They were grouped together by the Commission into seven 'clusters' of related research, as follows:

1. Oils: this cluster contains two projects dealing with vegetable oils and the processing of vegetable oils in the chemical industry.

2. Lignocellulose: this cluster contains a project concerned with lignin optimization in crop and industrial plants.

3. Carbohydrates: the seven projects in this cluster deal with the extraction of carbohydrates from different crops, carbohydrate processing by fermentation and sugar conversion processes, and studies of carbohydrates of particular economic interest.

4. Animals: this cluster contains five projects concerned with animal disease and nutrition, the production of bovine embryos and new methods of animal monitoring.

5. Biological pest control: there are five projects in this cluster of research, ranging from breeding for resistance to pests to the application of parasitic nematodes to the control of insects.

6. Crop production and storage: this is a heterogeneous cluster of projects involved with the production and storage of fresh fruits, the production of fertilizers and ways of improving *in vitro* plant cultures.

7. Proteins: this cluster includes projects on grass, silage and different types of peas. In addition, there is another project which concerns the use of biorefinery for whole crop harvesting.

The only other major political initiative to have fostered the non-food use of agricultural raw materials in the European Community was the amendment to the regulations allowing sugar and starch to be sold to the chemical industry at competitive prices (as discussed in Chapter 4). Certain products (those on a 'positive list') were allowed to benefit from a production refund which reduced the price of the raw material almost to world market level. The idea behind the premium of 7 ecu per 100 kg that the chemical industry was still expected to pay was that they should recognize the benefits of using an indigenous raw material. This has proved contentious, however, and efforts are still being made by the chemical industry to remove it.

The chemical industry maintained that, given competitive pricing, the volume of agricultural raw materials consumed for non-food purposes would increase as indicated in Table 5.1.

Table 5.1. Projected use of agricultural raw materials by the EC chemical industry/000 tonnes.

	1984–85	1991–92	1995–96
Sugar	62	180	500
Total starch	1260	2000	2600
Oils and fats	1700	2000	3500

Source: CEFIC (1985)

There are no market measures to encourage the non-food use of agricultural raw materials for energy purposes in the EC. However, in 1991 the Commission proposed a carbon tax of US$10 per barrel of oil or equivalent, from which renewable energy resources would be exempt. The Commission has also estimated that by 2010 8% of energy in the EC will come from renewable resources. For many non-food outlets the use of EC agricultural raw materials is simply not economically viable and there is insufficient political will to subsidize such non-food uses.

To date, the reform of the CAP has consisted of tinkering with the system rather than fundamental reform. A policy which is over 30 years old and has undergone repeated efforts to improve and amend it becomes riddled with inconsistencies, veritably 'creaking at the joints'. In the eyes of other industrial sectors which have to work with this policy, it is moribund. They prefer not to work within the framework of the CAP but rather to set up their production facilities outside the EC if possible. A bold and fundamental new approach is required, especially if the Community is to enlarge to include some of the East European states. It will be essential to separate the economic from the social aspects of an agricultural policy which attempts to legislate for countries of such diverse social and economic standing. Social aid has to be targeted to where it is needed; it is inappropriate to gear the whole policy to the least efficient. Efficiency and specialization should be encouraged by a truly economic policy which opens the way to new opportunities such as in the non-food sector.

Japan

Japanese agricultural policy objectives are stated in the Agricultural Basic Law of 1961 and bear remarkable similarity to those of agricultural policy in Europe (Australian Bureau of Agricultural and Resource Economics, 1988). There are five broad aims, which are almost identical

with those stated in the Treaty of Rome:
- to provide living standards comparable with those of off-farm households;
- to maintain the vitality of rural communities;
- to increase agricultural productivity;
- to provide stable prices to consumers;
- to ensure adequate food supplies.

The major problem facing Japanese agriculture is principally one of scale. 73% of farm households cultivate an agricultural area of less than one hectare. Rice is the most common crop, as it is a staple part of the Japanese diet and is of both cultural and religious significance. The nature of rice cultivation allows farmers to work part-time and augment their income from other sources. Thanks to the rapid growth of industrialization in Japan during the 1950s and 1960s, there are ample opportunities for farmers to work part-time in industry. Owing to the small scale of farming it is difficult for Japanese agriculture to compete on world markets with the result that a high degree of protectionism has been allowed to develop. Another key problem not dissimilar to that of Europe is the ageing of the farm population: according to the Maekawa report (1986) on the state of the Japanese economy submitted in 1986, 43% of farmers are more than 65 years old. Also, there is clearly a reluctance on the part of the younger generation to enter farming, even on a part-time basis.

The protectionist features of Japanese agricultural policy are quotas, tariffs, state trading and regulations that restrict the entry of agricultural imports. As in western Europe, the Japanese government practises price support for farmers which results in prices that average three times the world price of agricultural commodities. However, there has recently been internal pressure to reduce agricultural expenditure. To this end the producer price calculations have been altered for certain key commodities: wheat, barley and soyabeans. Efforts have also been made to increase the scale of farming by organizing agricultural cooperatives and allowing them to rent land from local farmers. This scheme has not met with much enthusiasm. This may in part be because the rents are not attractive enough for agricultural land, which commands a high price owing to the amount of agricultural protection. Further targets have been set to reduce the level of protection by 1992; these include reductions in the prices of rice, milk and beef, and the removal of import quotas on beef, oranges and certain processed foods.

The pressure for change is both internal and external. Japan's trading partners have exerted strong pressure for Japan to open its frontiers to international trade, and agriculture is no exception.

External pressure has come from the USA and Australia in particular. In common with other advanced industrial nations, Japan has pledged itself to reduce the level of agricultural support. Internal pressure for agricultural reform has come from the overall restructuring of the Japanese economy, which has placed greater emphasis on domestic demand than before. The prevailing political view is that Japanese people should benefit from four decades of industrial growth by the greater availability of consumable goods at affordable prices. Another catalyst for change is the financial drain on the economy of a relatively inefficient agricultural sector. In common with other countries there are popular demands to reduce agricultural spending. Finally, the ageing farm population gives cause for concern and is a further catalyst for major changes in farm policy.

In all these plans for reform, the non-food uses of agricultural raw materials do not figure highly. Because of its structure, Japanese agriculture will find it difficult to compete with other major agricultural producers for the production of agricultural raw materials for non-food purposes. There have been several initiatives to diversify land use on account of over-production. The first land diversification programme took place between 1971 and 1975 and was chiefly designed to divert land from rice production. However, little consideration was given to the alternative uses that could be made of the land. Diversion payments were offered to farmers for producing forage crops or perennials, and a slightly lower amount for taking land out of production altogether. Ironically, the amount of land planted to rice during this time actually increased. A second paddy field utilization programme was introduced in 1975–1978 with slightly more success, and there was diversification into fodder crops and food grains other than rice; but the payments for ceasing to produce were stopped on account of growth of urbanization on former agricultural land. A third paddy field re-orientation programme was introduced in 1978 and ran until 1986; this allowed crops other than those in over-supply to be grown on land withdrawn from rice production. This scheme was apparently more successful, as the target area for diversion was exceeded. The main crops grown on diverted land were wheat, barley and fodder crops. A fourth programme for diversification is currently in progress, which has the ambitious target of reducing the area under rice by a third.

Japan is still far from self-sufficient in wheat and barley production, so it is unlikely that they will be diverted into non-food uses. Furthermore, they can only be grown with considerable price support, which is now under threat through the GATT. Despite the strength of Japanese biotechnology industries it is unlikely that there will be a major synergy with its domestic agricultural production, which is not

Table 5.2. Production and use of rice in Japan 1975–1986.

Year	Area /ha × 10⁶	Production /kt	Consumption /kt	Import /kt	Export /kt
1975	2.76	13,165	11,964	29	2
1980	2.50	11,958	11,218	20	868
1986	2.30	11,647	10,370	41	0

Source: Ministry of Agriculture, Forestry and Fisheries (1987)

in a position to compete with other suppliers of agricultural raw materials for non-food purposes.

United States

Similar themes repeat themselves in the history of US farm policy. A fundamental 'Jeffersonian' belief in the virtue of small yeoman farmers underpins the protective policies employed by successive American administrations, whether Republican or Democrat. The principal goal of US farm policy is to maintain a prosperous, productive farm sector organized on the basis of family farms. The fundamental question facing US governments has been whether this should be achieved through the market place or through government intervention. Although there has been much discussion of applying free market forces to agriculture, most policies since the 1930s have seen at least some government intervention. The Republican Party has always been the stronger advocate of free market forces, whereas the Democrats have favoured income or price support coupled with supply controls. The choice between the two has usually been determined largely by world agricultural prices, but even Republican governments have tempered their enthusiasm for free market forces with the pragmatic use of farm income support. The change of administration in 1993 to Democratic control is not expected to have a major impact on the overall policy for agriculture in the United States. However, the Democrats, being innately more protectionist, may not drive the trade liberalization process as hard as did the Republicans before them.

Just as in Japan and Europe, there is political concern in the USA about the level of income obtainable from farming compared with other sectors of the economy. This is expressed in terms of income parity as a goal for farm policy. In practice it has been achieved by a variety of methods: supported farm prices or price guarantees, as well as loan rates and direct income support. However, pressure has grown from an increasingly urbanized electorate against the treatment of agriculture as a special case and in favour of reducing budget expenditure on agriculture. This pressure was successfully mitigated during the 1970s on account of expanding export markets for American agricultural products, particularly grain. During the Nixon administration the commencement of sales of wheat to the USSR enabled the President to buffer some of the pressure to reduce agricultural support. The relatively easy passage of the US Farm Bill in 1981 reflected the market expansion made possible by foreign and domestic economic growth. This short period of unrestricted growth only put off the evil day when the US administration had to face the need to tackle the choice between continuing to protect its farmers or allowing free market forces to play.

One interesting feature which recurs in the history of American farm policy is the concept of taking land out of food production. Large-scale land retirement schemes were first discussed in the 1930s. These have been repeatedly discussed and indeed practised as a temporary means of bringing supply back into balance with demand. The set-aside tool has not been effective in reducing the overall volume of agricultural production and it is certainly not seen as an effective means for promoting the non-food use of agricultural raw materials. Unlike Europe and Japan, the United States does have a major programme for the non-food use of agricultural raw materials for energy.

It is significant that it was not so much the oil crisis of 1973 which prompted the US government to turn its attention to the possibility of producing energy from agricultural raw materials as the crisis in agriculture. By the mid-1970s agriculture had been grappling with the combination of over-production and depressed prices for some time. The agricultural lobby then launched a campaign to convert excess grain into liquid fuel. The US government responded by passing legislation in 1977 exempting 'gasohol' from the $0.04 per gallon Federal excise tax on petrol. Gasohol was defined as a mixture of at least 10% ethanol and 90% petrol. Since each gallon of ethanol can be used to produce 10 gallons of gasohol, the $0.04 per gallon exemption translates into a tax benefit of $0.40 per gallon, or $16.80 per barrel. In essence, this benefit allows fermentation ethanol to be priced $0.40 per gallon above the price of high-octane fuel and such additives as toluene, benzene and xylene which are also used as octane boosters. This tax relief was originally scheduled to expire in 1974 but was extended until 1992. During this time, it was increased twice, first to $0.05 and then to $0.06 per gallon, but it has since been pared back to $0.054 per gallon. This policy has been extended to the year 2000.

In addition to this direct tax benefit for ethanol, the US government introduced incentives aimed at increasing the production of fermentation ethanol. In particular, a double tax credit of 20% was made available for investment in ethanol production facilities. Individual states producing the commodity have also offered incentives for bioethanol production. Several maize-producing states such as Colorado, Kansas, Iowa, Illinois, Missouri, Michigan and Wisconsin, have exempted gasohol from state excise taxes, which are normally $0.05 per gallon, providing the raw materials are produced within the state. In a state like Wisconsin, ethanol attracts in effect a $1 per gallon ($42 per barrel) tax exemption. This is a remarkable incentive when compared with current crude oil prices in the region of $18–20 per barrel.

In order to appreciate why a tax incentive of this order should be paid, it is necessary to understand the agropolitical context. Maize production in 1982 was 70% higher than in 1967, and the surplus stored from one year to the next doubled between 1977 and 1982. In the mid-1970s the USA was forced to sell large quantities of grain to the Soviet Union, in part because it simply could not be stored. Owing to the extensive over-production the real price of maize had fallen steeply.

Another factor taken into consideration was the amount of subsidy being paid to farmers not to aggravate the existing grain surplus. This 'no grow' payment was set at $1 per bushel that the land could have produced. Since one bushel of maize will yield approximately 2.5 gallons of ethanol, it was considered preferable to pay a $0.40 per gallon ethanol tax exemption rather than a $1 per bushel non-production subsidy.

As a result of this incentive it is not surprising that ethanol production took off rapidly and on a large scale (Hacking, 1986). US ethanol production capacity now exceeds 500 million gallons annually. Gasohol accounts for 3–4% of total petrol sales. 120 million bushels of maize are needed to produce this quantity of ethanol, out of a total maize production of 8000 million bushels. If all US maize production were converted to ethanol, it could meet about 20% of all US petrol needs.

The original idea of the US government was to promote the construction of small, even on-farm plants, each producing up to 1 million gallons per year. With these incentives many medium-scale operations were set up by wet and dry maize-milling firms. Many plants failed even to start up, and by 1982 there were only 50–100 small plants producing 15–20 million gallons of ethanol per year. There were various 'teething troubles' when these plants started, notably corrosion problems, inadequate cleaning leading to infection, and killing of the yeast through ignorance about temperature and pH requirements.

By contrast, the large-scale operators have been extremely successful. Over 90% of the ethanol production capacity in the US is owned by nine major producers. One producer alone, Archer Daniels Midland, accounts for 50% of production. It is a major starch-producing company with only 20% of its total starch production used in ethanol manufacture. The six largest ethanol plants have a capacity of over 10 million gallons per year and benefit from economies of scale. In the opinion of those who have personal experience of running these large-scale bioethanol projects, only these large wet milling operations are able to yield a return on investment sufficient to reward the risks.

Concluding this section on the appropriateness of agricultural policy for promoting non-food uses, it can be observed that important elements of agricultural reform are being driven by social policy rather than economic considerations. Indeed this is the fundamental weakness of post-war agricultural policies worldwide: they have outlived their usefulness as tools for postwar economic reconstruction and now act as hindrances to economic growth. As far as the non-food uses of agricultural raw materials are concerned, the shying away from market forces has handicapped agriculture's ability to compete with oil. The failure to hive off the social aspects of agricultural policy has impeded those who are able to compete from taking advantage of new markets for their produce.

OTHER POLICIES AFFECTING THE PROGRESS OF NON-FOOD USES

The US Office of Technology Assessment (OTA, 1984) identified a range of factors relating to the legal system and various government policies which help or hinder the commercialization of a new technology:

- financing and tax incentives;
- government funding for basic and applied research;
- training and personnel ability;
- health, safety and environmental regulations;
- intellectual property law;
- university–industry technology transfer;
- international technology transfer and trade.

These factors are worthy of examination individually and provide a convenient framework for assessing the policies which affect the development of non-food uses.

Financing and Tax Incentives for Non-food Uses

Manufacture of a non-food product from agricultural raw materials is undertaken either by an existing company wishing to expand its range of products or by a new company; both require some form of financing, and both are affected by tax laws for business. More new companies have been established in the United States to commercialize new technologies in the non-food sector than in Japan and Europe, where venture capital is less readily available (Price Waterhouse, 1992). In Japan and Europe commercial efforts in the non-food domain are led by established companies. The large financial assets of such firms have no need for external funds for research and development (R & D) in areas such as biotechnology. It is also easier for companies such as these to obtain debt funding, which is not usually available to small new firms starting up in this area.

Tax Incentives

Tax provisions vary from country to country, and it is those concerning corporate taxation, the treatment of small businesses, capital formation and R & D expenditure which affect the competitiveness of the individual firms. Some countries rely on tax provisions more than others to stimulate capital formation. The USA and the United Kingdom, for example, rely more on tax incentives than do Germany or France, where more grants and subsidies are available. Japan targets particular industries and employs both grants and tax incentives. Some countries set corporation tax (CT) rates as low as 3% to attract foreign investment. When compared to the equivalent rate of CT in the USA, for example – currently 46% – this looks attractive but in practice investors will weigh up other factors which offset such apparent attractions before making their decisions.

Tax incentives can be used to encourage the development of non-food uses both up- and down-stream. The treatment of income from the sale of licence technology differs from one country to another. For example, the proceeds from the sale of patents in the USA is treated as long-term capital gains and royalties are treated as ordinary income. In Japan, both proceeds and royalties are treated as ordinary income. European practice varies from one country to another with some startling contrasts; for example, the sales of patent rights in France are taxable at the reduced 15% long-term capital gains rate, whereas in the UK CT rates as high as 52% apply, depending on the size of the company. The USA has the most favourable tax treatment for new and expanding businesses in the non-food sector. Recognizing its competitive disadvantage, the European Commission is trying to develop policies which encourage entrepreneurs and harmonize the situation between member states, but there are both fiscal and cultural barriers to overcome.

Government Funding of Basic and Applied Research

The development of non-food uses, particularly the introduction of new processes and products, is related to the amount and quality of research carried out in each country. Some governments have a more generous attitude to state funding of research than others. In Japan funding for research into non-food uses is divided among the Ministries of Trade and Industry, the Science and Technology Agency, and the Ministry of Agriculture, Forestry and Fisheries. Japan has shown a particular interest in bioprocess engineering and a commitment to biotechnology as offering the 'next generation' of products for the non-food use of agricultural raw materials (Saxonhouse, 1983). The agency for Industrial Science and Technology has sponsored a large amount of work in universities and research institutes, including studies of the industrial use of enzymes and the utilization of biomass. Biotechnological research alone accounted for expenditure of $60 million in the mid-1980s.

In the EC, there are two tiers of research: Community level and national level. At Community level there is a Directorate General designated for Science, Research and Technology (DG XII), but its funds represent only a small part of the overall Community budget. Research funding is usually made conditional upon two factors; firstly, international cooperation where the projects must involve more than one member state; secondly, shared funding with national governments. The condition of international cooperation is a good ideal but often proves unwieldy and is hindered by the language barriers in Europe. National governments are sometimes reluctant to put up their share of the funding of an EC project. Research expenditure varies considerably within the Community from one member state to another. The Commission has recently designated 100 million ecu to be spent on research into non-food uses of agricultural materials out of a total agricultural research budget of 500 million ecu.

In Germany, there is a specific Ministry for Research and Technology with a sizable budget for research, but other ministries such as that for agriculture and industry have their own research funds. The German government operates one of the best and most ambitious institutions for the advancement of biotechnology in the world: the Society for Biotechnological Research, with an annual budget equivalent to $13 million.

In the United Kingdom, several government-sponsored research centres are looking into the scope for non-food uses. These are in the process of rationalization and in some cases privatization or at least taking on significant industrial sponsorship. A successful example of this is Celltech, which was originally founded by the National

Enterprise Board in 1980 but is now part-owned by well-known financial institutions such as the Midland Bank and the Prudential Insurance Company. It is a good example of the British government policy of involving industry and private finance in capitalizing on research undertaken in the public sector (Advisory Council for Applied Research and Development and the Royal Society, 1980). Agricultural Genetics is a company of similar design which is commercializing the results of research undertaken by the Agricultural and Food Research Council (AFRC). There is a basic principle guiding UK government research that it should be precompetitive. This has tended to prevent the UK from taking advantage of EC funding for demonstration-scale projects, as these are considered too close to the market place.

France, unlike Germany, does not have a centralized research society but rather a number of research centres reflecting the classic divisions between ministries such as the National Institute for Agricultural Research and the National Centre for Scientific Research. There is also the famous Institut Pasteur, which is jointly funded by government and industry, and which carries out some work into non-food uses.

The US Government has a far greater commitment to the funding of basic research than to applied research. It spends more than any other nation in the world on basic research into biotechnology. In the USA, basic research is funded by government departments such as the US Department of Agriculture or the Department of Energy. There is some duplication between them; for example, the Department of Energy carries out research on renewable energy and conservation. There are also several government-funded research institutes, such as the National Science Foundation. As regards applied research, the US Congress has made a strong commitment to a reasonable share of Federal research money going to small business and this may be another reason why commercial applications of non-food research are more common amongst small new businesses in the USA than elsewhere.

Personnel Availability and Training

A vital factor in the competitiveness of a country in the development of new markets for non-food uses of agricultural raw materials is the availability of adequately trained personnel. This depends on the quality of training available in each country and on each government's commitment to education and training. The success of individual companies in their attempts to commercialize new non-food products will depend on the quality of the human resources they have to undertake the tasks leading up to marketing the product. Since the new generation of non-food products often employs new technology such as biotechnology, commercial development is highly dependent on skilled personnel. The types of skill required heavily reflect the new techniques of genetic engineering, with specialists such as bioprocess engineers, enzymologists and cell culture specialists, both those who work at laboratory-level basic research and those involved in scaling up.

In Japan there was a lack of government support for basic research in universities, and a consequent lack of skilled personnel in the biological sciences during the 1980s, which companies sought to rectify with in-house training programmes. Another inherent problem which had to be overcome was the comparative lack of personnel mobility between companies and overseas. The success of these initiatives means that more bioprocess engineers are working in Japan than in any other competing country in the non-food sector. One of the historical strengths of Japan stems from the fact that after the Second World War, when most chemical companies were switching to petrochemical feedstocks, Japanese chemical companies retained some biomass feedstocks and have thus come to dominate certain non-food markets.

In Europe, there is a general concern about a 'brain drain' of skilled personnel to other countries, the USA in particular. This is a more serious problem for some member states than for others; the exodus of qualified personnel is greater from the UK than from Germany. This is particularly tragic in view of the quality of education and training in a country like the UK which invested heavily in excellence in this field in the life sciences during the 1950s and 1960s, and then failed to reap the full benefit. Even during the mid-1970s spending on health-related R & D was twice that of its competitors in Japan, France and Germany. The main reason for the brain drain is the lack of posts and the lack of competitive salaries.

The USA has the largest number of specialists in genetic manipulation in the world. This is one of the reasons why so many small new businesses sprang up to commercialize biotechnological research. Few companies experience shortages in personnel trained in the basic biological sciences. The USA also benefits from attracting trained personnel from overseas. If there is any area of weakness it is in the area of plant molecular biology, resulting partly from a lack of government funding in this area and partly from the industrial switch away from biofeedstocks to petrofeedstocks after the Second World War.

Health, Safety and Environmental Regulations

The development of non-food outlets can be impeded by the regulatory framework in force. Particularly because the non-food product may be novel or employ novel technology such as genetic engineering in its manufacture, it is almost bound to be the subject of health, safety or environmental regulation. When the techniques of genetic engineering were first devised, there was great concern about the risks of introducing genetically engineered organisms into the environment. Now that scientists know more about molecular genetics, however, they can gauge the accuracy of perceived risks and risk assessment has become more accurate.

The cost and time involved in complying with the regulatory framework is the price society pays for safety. The regulatory framework varies from country to country. Those countries that have the fewest restrictions and uncertainties in their regulatory framework affecting non-food products will have a competitive advantage. There are several levels of regulation affecting different stages of non-food production: some regulations pertain to the processes of production, some to the product made and some to the health and safety of workers involved in the production. Again all these vary from country to country.

The use of the techniques of genetic engineering in non-food products made from agricultural raw materials is subject to international guidelines set up as a result of concern about the potential risks of recombinant DNA research in the 1970s. The most stringent guidelines in force are in Japan, where companies have to seek government authorization for large-scale cell cultures. Under US guidelines large-scale work need be reviewed only by a biosafety committee; a similar situation exists in the UK, where large-scale research is treated on a case-by-case basis by the Genetic Manipulation Advisory Group. Most countries assess risk according to the source of the DNA and the pathogenicity of the host–vector system. The recommendation of the EC is that notice of experiments should be given to the central authority in each member state before work using genetic manipulation begins. Some member states allow certain types of work to be carried out once the appropriate health and safety authority has checked the ability to contain risk.

Regulatory control of non-food products is subject to the same constraints as any new products entering the market place, regardless of whether or not a novel technique has been used in their production. Before launching a new non-food product the producer has to weigh up the following questions:

- How much time and effort will it take to obtain approval for the new product?
- What is the usual cost for securing regulatory approval?
- What are the import and export restrictions on the product?
- What sort of test data do the regulatory authorities require?

The responses to these questions will affect the choice of country where the new product is launched. In the USA there are three key regulatory authorities: the Food and Drug Administration (FDA), the US Department of Agriculture (USDA) and the Environmental Protection Agency (EPA). The FDA regulates food, food additives and drugs, so non-food products such as vitamins and antibiotics would have to obtain FDA approval. Those made using novel techniques such as recombinant DNA technology certainly have to go through stringent new product approval procedures. The USDA shares a common procedure with the FDA and concentrates its own regulatory efforts on products destined for use by animals such as animal vaccines, rather than those for human consumption which are the jurisdiction of the FDA. The EPA covers a broad spectrum of products including chemicals, herbicides and pesticides. It operates under the Toxic Substances Control Act, which authorizes it to identify and evaluate 'chemical substances' and then regulate their production, distribution and use. There is a premarketing procedure under which the EPA reviews data on safety and, providing it will not generally cause adverse effects to the environment, the product is then registered.

The EC is in the process of harmonizing the testing and approval procedures for new products between member states as part of its moves towards an integrated single market. However, there are still national differences within the Community at this stage. There is concern amongst individual governments that an EC-wide standard for testing new products might be set to the lowest common denominator, which would then be unacceptable to member states with a higher standard of testing. For example, in Germany a pharmaceutical manufacturer must obtain individual marketing authorizations to distribute its product as well as manufacturing authorizations for each of its production plants. Import restrictions are also extremely strict: an importer is required to demonstrate that the product's foreign manufacturer has the equivalent of a manufacturing licence. Restrictions on exports are slightly lighter: the exporter does not need a manufacturing licence.

In the United Kingdom a similar system of product and manufacturing licences applies. In addition, however, products that are produced by biotechnological means are licensed on a batch release basis. This means that each batch of the product is subject to certain tests which are then submitted to the National Institute of Biological Standards and Controls. Importers have to obtain a product licence and the licensing authority is permitted to inspect the manufacturer's premises. Export licences are required, but these consider only the quality of the product, not its safety or efficacy.

In France, marketing licences are granted for a period of five years and can be renewed providing no modifications have been made to the original process. Imports require a recognized marketing licence from the country of origin.

In Japan, as in Germany, separate licences are required for the product as well as the premises of manufacture. Manufacturing licences are for three years' duration, providing there is no major modification to the process. Product licences are valid for six years except in the case of a novel product, the safety of which may be re-assessed after four years.

In conclusion, it can be said that health, safety and environmental regulations can affect the cost, time and financial risks of getting a product to the market place. Where genetic engineering is employed in the production of non-food products there are variations in the regulations from one country to another. Most of these concern the handling of products making use of recombinant DNA technology. They are essentially voluntary, however, and directed primarily at research. The USA has the least restrictive guidelines of developed competing countries. The vast majority of experiments using recombinant DNA can be done at the lowest containment levels; prior approval is only sought for a limited category of experiments. Japan has the most restrictive guidelines for experiments using recombinant DNA for which prior approval has to be obtained. In the European Community, the United Kingdom and France have the least restrictive regulations and Germany the most restrictive. There are no significant differences in the health and safety regulations affecting workers in this area.

Intellectual Property Law

The capacity to protect an invention, novel product or technique is considered an important feature of any decision by a private company to invest in its production. The laws that are used to protect such things are known as intellectual property laws and cover such instruments as patents, trade secrets laws and breeders' rights. The use of biotechnology in the non-food use of agricultural raw materials necessitates an examination of the intellectual property laws applying in each country because biotechnology gives rise to a vast array of new inventions. These laws must cover both the products and the processes if they are to be effective. Variations between them will affect the competitiveness of one country *vis-à-vis* another.

The objective of patent law is to grant inventors a limited property right over their inventions. In the USA, a patent gives the inventor the right to exclude all others from making, selling or using the invention without the inventor's consent for a period of 17 years. In return the inventor must make a full public disclosure of the invention. Western Europe has an international treaty on patents known as the European Patents Convention (EPC). The EPC was set up in 1977 and has established a legal system for granting European patents through a single European patent office. The EPC system exists in parallel with national laws for granting patents; the ultimate goal is for the EPC to be adopted into national law and for a single European patent to exist. Under the EPC, patentable subject matter is anything susceptible to being applied industrially. This broad definition belies some very restrictive limitations. So, for example, scientific theories, diagnostic methods of treating humans and essentially biological processes, animals, and plant and animal varieties are not patentable. This means that in Europe inventors have to seek their protection from some other type of intellectual property law.

In Japan, there is a similarly broad definition for patentable subject matter: 'any person who has made an invention which can be utilized in industry can apply for a patent', and these provisions have been extended to microbiological inventions. The disclosure requirements are similar to those in the USA. Under the Japanese Patent Act of 1976 an application for a patent has to be accompanied by a detailed explanation of the invention. USA patent law allows greater dissemination of scientific information about an invention than that in either Europe or Japan, which probably places the American inventor at a slight competitive advantage.

European inventors of microbiological processes or products cannot obtain adequate protection for their inventions under patent law and so are forced to resort to laws on trade secrets or breeders' rights for protection. There is no common standard for trade secrets in Europe, however. Germany has the strongest statutory system for protecting proprietary information and its courts are consistent in enforcing the statutes. The UK is slightly more relaxed, emphasizing confidentiality over secrecy in the use of such information. France is more relaxed again. Japan is even further behind Europe in the protection of proprietary information, but as it changes from a technology-importing country to a technology-generating country improvements in trade secrets law can be expected.

Even the laws on breeders' rights afford better protection in the USA than elsewhere. Ownership rights of plant material are covered by two laws: the Plant Patent Act of 1930 and the Plant Variety

Protection Act of 1970. The latter provides a patent-like protection for 17 years to new, distinct, uniform and stable varieties of plants which can be reproduced sexually. In Europe and Japan, protection of breeders' rights is in the form of protection certificates rather than patents. Thus the protection provided in these countries is relatively limited compared to that in the USA. There is strong political opposition to the application of patents to plant varieties in Europe, but Japan may be moving in that direction. In the EC there has been discussion for some time of a new Directive which would amend existing legislation that exempts plant and animal varieties from patenting, but it is unlikely to be accepted in its present form.

There have been efforts to coordinate international plant variety protection. These culminated in a treaty drawn up by the International Union for the Protection of New Varieties (UPOV), which was signed for the first time in 1978. It is reviewed and updated at biennial congresses, the most recent of which took place in March 1991. The signatories of the treaty recognize the difficulty of protecting plant material, and its wording is sufficiently loose to allow double protection under the Treaty and under US patent law, and also under the European system of breeders' rights.

Industry/Technology Transfer

In Japan, the interaction between academia and industry in the non-food domain is extremely limited. This stems from certain traditions in Japanese universities, where there is a strong distinction between basic and applied research. Japanese professors of basic research in disciplines such as biology pride themselves on their independence from industry. By contrast professors in applied sciences have on-going contacts with industry and almost a sense of moral obligation to provide these industries with their future employees. Despite this the level of information exchange is still not as high as in the West. Professors are restricted from taking on other positions in parallel with their university work. All donations to their work must be made through formal university channels. About 10% of all university funding in Japan comes from industrial sources.

In Europe it is impossible to generalize about the relationships between industry and academia because these vary from country to country. Germany has a particular problem because it is the individual Länder which regulate research at their own universities. Furthermore they have to compete with privately owned research institutes, such as the Max Planck Institutes, which attract private funds away from the universities. Another hindrance to technology transfer is the lack of cooperation between disciplines, particularly the traditional separation of the technical faculties from their arts and science counterparts. These obstacles have actually led large German companies, such as Hoechst Technology, to invest in research facilities outside the country.

The United Kingdom is also weak in this area. The gap in the area of strategic applied research is known as the 'predevelopmental gap'. The British Government has recognized this weakness, particularly the failure of the National Enterprise Board which was set up to foster university/industry relationships. To encourage direct links between academia and industry the Cooperative Research Grant scheme was set up under the Science and Economic Research Council. The SERC will support university research providing a company makes substantial contributions in efforts, materials and expertise. Another means of transferring technology is through the role of consultants which academics in Britain are free to play. Britain has an excellent research base but difficulty in translating this expertise to industry, a fact not helped by the fluctuating economic performance of this country.

The French universities' main objective is education; therefore most highly regarded research is carried out in specialist institutes. Despite there being no formal constraints upon relationships between academics and industry, these have not flourished. This is probably because opinion is divided over whether such relationships should be encouraged. There is a National Agency for bringing value to research which is designed to encourage the transfer of public research results to industry. Large French companies such as Elf-Aquitaine and Rhône Poulenc have taken advantage of this and, significantly, have located new research facilities near certain universities with traditional strength in the biological sciences. The French government has tried to improve contacts between the research institutes and industry through a committee for the organization of strategic industries, but has met with only limited success.

The USA probably has the world's most effective and dynamic relationships between academia and industry. This is chiefly because of the openness of the university system and the freedom of faculty to pursue research. There are many mechanisms by which American academics can pursue their relationships with industry: dissemination of publications, interprofessional meetings, consulting arrangements, contract research, cooperative research arrangements and the setting up of industrially funded institutes within universities. All these mechanisms help to ensure the transfer of technology and the diffusion of knowledge.

Trade

Barriers to trade of non-food products include export controls, exchange and investment controls, non-tariff barriers and price regulation. In Japan, export controls combine trade concerns with defence and foreign policy objectives. The list of controlled items includes blood derivatives, fertilizers and bacterial agents for military use. Export controls in Germany and the United Kingdom are much less restrictive, being confined chiefly to 'implements of war'. The USA's export controls are more restrictive than those of its competitors. Foreign exchange and investment control can have the effect of restricting the importation of foreign technology, capital or products, thus giving local firms an advantage in their domestic markets. The USA and most European countries have no significant formal exchange or investment control laws. Japan has an investment and exchange control mechanism which does affect technology transfer and foreign investment. Japan is also the most restrictive regarding technical assistance and licensing arrangements between foreign parties and Japanese companies. The Japanese government has the power to screen investments before a foreign investor can conduct a transaction. Certain companies are under the specific control of the Finance Minister as regards foreign ownership. Whilst a limited amount of foreign investment is permitted, the waiting period for authorization is at least a nuisance to the potential foreign investor, making the Japanese market notoriously difficult to penetrate.

Non-tariff barriers to trade include any government intervention affecting competition between imported and domestic goods. These include:

- subsidies;
- price regulation;
- health and safety standards and certification;
- Customs classification.

Subsidies are a hotly disputed area of trade law because they can provide an unfair competitive advantage to domestic producers. Non-food products are directly affected by the Subsidies Code established by the Tokyo Round (1973–1978) of the GATT negotiations, at which it was agreed that no industrial product should be allowed an export subsidy. Domestic subsidies are allowed providing they do not threaten injury to another trading partner's industry. So, for example, products that are listed as eligible for a production subsidy under the EC's production refund system for non-food products made from sugar and starch are at present allowed to continue receiving this support as they are produced only in small quantity and not usually exported. If, however, a non-food product on this list begins to be produced in large quantity (as has happened with biodegradable plastic wrapping), the situation might have to be reviewed. To offset any possible market distortion, the Subsidies Code allows national governments to impose countervailing duties on subsidized imports.

Price regulation and government procurement can also have a distorting effect upon trade. Although the motive may be innocent, such as the prevention of exploitation of the consumer or the control of inflation, price regulation can give local producers an unfair advantage. The GATT regulation requires the products of its signatories to be treated on equal terms. Particular problems arise in the area of health care, where governments do tend to intervene to contain the cost of treatment. Some governments set an official selling price for drugs, with a built-in profit margin for the drugs companies. The degree of profit may reflect the company's commitment to undertake research locally, and such a condition would be in breach of the GATT. There are distinct inequalities between trading partners in this area. In countries like Germany and Italy, all drugs are dispensed through the pharmacy and such price regulation is very tight. On the other hand, in the USA 'over the counter' sales of drugs in non-specialized stores are far greater – and thus there is a greater play of market forces. In Japan, almost all drugs are dispensed by physicians, whose drug lists and mark-ups are regulated by the national health insurance system.

The GATT regulation allows governments to buy products for their own consumption and target their procurement to favour local suppliers. This allows the European Community to practise its intervention buying mechanism for agricultural produce. As discussed earlier, the agricultural policy of the EC places heavy emphasis upon security of supply; part of the system involves buying in of intervention stocks of food if prices fall below a predetermined level. This provision is questionable in trade terms as it gives domestic producers an unfair advantage over their competitors and prevents the price of food finding its 'natural' level as a result of market forces. Since the intervention purchases are usually made as a result of falling prices when demand exceeds supply, the stocks usually end up having to be exported, at considerable loss to the taxpayer.

Customs classification can prove to be an obstacle to trade. Some nations are guilty of protectionism in making the procedure for importing new products employing novel technology particularly onerous on various false pretexts. Furthermore, for a foreign manufacturer registration and approval of a product may involve a certification process, resulting in an inevitable leakage of technology.

In conclusion it may be said that the significant trade barriers to the international technology transfer of non-food products are non-tariff ones. These are within the scope of GATT rules. The Standards Code prohibits the signatories from discriminating against imports in their standards and certification systems and the Subsidies Code prohibits certain forms of market-distorting subsidy. Unfortunately, whilst the trading partners pay lip-service to freeing their markets of these distorting features there is a clear reluctance to remove them completely.

A BLUEPRINT POLICY FOR NON-FOOD USES

This chapter, and indeed this book, cannot be concluded without trying to draw together the lessons learned from examining the individual areas affecting non-food uses: the technologies, the economics and the policies, and attempting to suggest how these might be improved. One fundamental lesson to be learned is that existing agricultural policies are often inappropriate for the advancement of non-food uses. They are orientated to food production and try to embrace too many social features for them to be an effective policy framework for non-food uses. It is insufficient to 'tinker' with existing regimes for food production in the belief that this will provide the right policy basis for a wholly new production area. As long as non-food uses are considered only in the context of a food policy, the policy framework will be imperfect.

A good illustration of this is the attempt to get fuel uses of agricultural raw materials off the ground by diversifying the use of land set-aside from food production. The payments to farmers for taking land out of production bear no relation to the economics of the industrial uses to which the land could be put. Furthermore, it is questionable how long it is possible to maintain public support for paying farmers not to produce when other industrial sectors do not obtain compensation of this kind. Set-aside may be sustainable on a small scale, but idling vast tracts of agricultural land may prove difficult to sustain politically. Evidence for this is provided by the USA, where in 1978 the cost of idling land rose to $7.9 million (Paarlberg, 1980). This raises the question of who will invest in a commercial process dependent on a subsidy such as this which may prove to be politically vulnerable.

A better option might be a decision to hive off non-food policy from being an annexe to food production and to treat the subject on its own merits. So, for example, the energy use of agricultural raw materials would be treated as a part of energy policy and the chemical use of agricultural raw materials as part of industrial policy. This places the policy emphasis back on the demand side, which is likely to produce a less convoluted solution than the adaptation of food policy to the demands of unrelated industrial sectors.

This does not mean that only a short-run economic view should be taken of non-food uses in non-agricultural sectors. On the contrary, the long-run economics, social and environmental considerations and domestic and international political considerations are just as important. But they can be assessed in the context of the sector agriculture will be supplying and not purely driven by the situation in agriculture itself. The former is dynamic, the latter contrived. Governments should assess carefully the macroeconomic factors affecting these new markets for non-food uses before deciding what type of subsidy, if any, should be paid to promote non-food uses. A model for these, proposed by Buckwell and Young (1986), included:

- future oil prices;
- the pace of technological change;
- which biomass route to take;
- differences between countries;
- implications for trade flows;
- externalities (including environmental considerations).

These are all subjects treated in some detail earlier in this book. The range of oil price forecasts indicate an increase in the long term. Biotechnology is continually delivering new possibilities for agricultural raw materials to be used for non-food uses. This technology is far from being mature. Chapter 2 indicates the range of both established commercial crops and novel crops which could be grown for non-food purposes. This chapter has indicated the political, cultural and economic differences between the countries competing for non-food uses, which finds the European Community lagging behind its partners on account of its lack of harmonization in legislation affecting this area. There is extensive treatment of the externalities, particularly the environmental considerations in respect of non-food uses. The environmental benefits of non-food uses outweigh the disadvantages, providing the raw materials and products are produced with regard to the need to protect the environment.

No country has a comprehensive policy on non-food uses and there is a good case for some international cooperation in the design of a policy suited to this sector. Owing to the globalization of markets, it is no longer practical for an individual country to make policy in a vacuum. Conflicts and agreements profoundly influence world trade (Josling, 1985). A willingness to cooperate as expressed in the GATT negotiations to liberalize world trade could be built upon to create an international non-food policy and thus avoid some of the mistakes of narrow protectionist policies in the past.

REFERENCES

Advisory Council for Applied Research and Development and the Royal Society (1980) *The Spinks Report.* HMSO, London.

Australian Bureau of Agricultural and Resource Economics (1988) *Japanese Agricultural Policies.* Australian Government Publishing Service, Canberra, Globe Press.

Buckwell, A. and Young, N. (1986) The implications of using biomass for energy. In: Sourie, J.C. and Killen, L. (eds) *Biomass: Recent Economic Studies.* Elsevier, Amsterdam.

EC Commission (1985) *Perspectives for the Common Agricultural Policy.* Brussels.

EC Commission (1986) Towards a market-driven agriculture. Discussion paper from the Concertation Unit for Biotechnology in Europe, Brussels.

EC Commission (1987) The agro–chemo–energy complex. Paper from the Forecasting and Assessment in Science and Technology team, Brussels.

EC Commission (1988) *The Common Agricultural Policy: Four Years of Reform.* Brussels.

European Council of Chemical Manufacturers Federation (CEFIC) (1985) *The Use of Agricultural Raw Materials in the European Chemical Industry.* CEFIC, Brussels.

Hacking, A. (1986) *Economic Aspects of Biotechnology.* Cambridge Studies in Biotechnology, Cambridge University Press, Cambridge.

Josling, T. (1985) Commentary on Tangemann, S. Repercussions of US agricultural policy for the EC. In: Gardner, B. (ed.) *US Agricultural Policy.* American Enterprise Institute for Public Policy, Washington.

Mackawa, H. (1986) The report of the advisory group on economic and structural adjustment for international harmony. Submitted to Prime Minister, Mr Y. Nakasona, 23/4/86, Tokyo.

MAFF (Ministry of Agriculture, Forestry and Fisheries) (1987) *Food Balance Sheets.* Tokyo.

OTA (Office of Technology Assessment) (1984) *Commercial Biotechnology – an international analysis.* Pergamon Press, New York.

Paarlberg, D. (1980) *Farms and Food Policy.* University of Nebraska Press, Lincoln, Nebraska.

Price Waterhouse (1992) *Doing Business in Germany, Doing Business in France, Doing Business in the United Kingdom, Information Guides.* Price Waterhouse World Firm Ltd, London.

Saxonhouse, G. (1983) *Biotechnology in Japan.* Contract report prepared for the Office of Technology Assessment, US congress.

6

Conclusion

The non-food uses of agricultural raw materials afford an opportunity for farmers to exploit the full potential of biomass. At different times in human history, agricultural raw materials have been used for both food and non-food purposes. The present neglect of the use of agricultural raw materials for non-food purposes has resulted from their substitution by alternative raw materials such as fossil fuels. But both the finite nature of fossil fuel resources and the fact that burning them has negative environmental consequences have led to fresh interest in renewable agricultural raw materials. The extent and the timing of the growth in the use of agricultural raw materials will depend on various economic, technical and political factors. Of these, it seems the last-named will play the most important role but it is also the most difficult to predict.

The technical obstacles to the growth of non-food uses are not on the whole insurmountable. The rapid progress of new technologies such as biotechnology have provided a solution to a whole range of problems in using agricultural raw materials for non-food purposes. Genetic engineering has allowed the modification of living organisms to adapt them better to specific industrial uses. Some of the remaining difficulties, such as engineering the capacity for cereals to fix nitrogen, are complex but not impossible to overcome. There are some practical problems associated with the handling of agricultural raw materials, notably on account of their bulk and their perishable nature, but these too can be addressed with modern methods of harvesting, transportation and storage albeit at some cost. It is unlikely to be the technical difficulties which prevent the growth of non-food uses.

A difficulty that will be harder to overcome is the lack of economic viability for some non-food uses, particularly where agricultural raw materials have to compete with oil. This is certainly the case for energy applications of agricultural raw materials. However, the present situation is not static and the underlying economic trend is for the cheapening of agricultural raw materials relative to hydrocarbons. The price of oil is, at the time of writing, depressed through overproduction and does not reflect its true environmental cost. If these factors were reflected in the price, the break-even point for substituting hydrocarbons by agricultural raw materials would come sooner. As it is, the improving economic viability of agricultural raw materials is in large part due to the reduction in agricultural prices in real terms as levels of support have been reduced worldwide. The recent Uruguay Round of the GATT negotiations has brought further progress towards a reduction in agricultural protectionism and, notably, artificially supported agricultural prices.

It is possible to imagine an interim or transitional scenario where a two-tier pricing system for different uses of agricultural raw materials exists. Agricultural prices for food uses are managed in many economies around the world in recognition of the strategic importance of food and farming. Whilst agricultural support may be gradually reduced it may not be completely removed for some time. However, in order for agricultural commodities to be able to compete with other raw materials such as hydrocarbons it is necessary to supply them as near to world price as possible. Therefore it can be anticipated that a two-tier pricing system may operate for a while where food uses enjoy a degree of agricultural support whilst non-food uses do not. This inevitably implies a degree of cross-subsidization which is only possible whilst non-food uses remain the minority use of agricultural raw materials. Food uses are likely to remain the predominant purpose of agricultural production for quite some time, particularly with the acceleration of doubling times in world population growth.

The underlying dynamic of population growth adds weight to the case for greater interdependence in agricultural production. The increasing demand for both food and energy from a growing population will require a willingness to depend on the comparative advantage of some areas to produce these commodities over others if limits to use are to be extended over time. At present there is an unnecessary degree of duplication amongst the world's agricultural producing nations which does not result in a rational use of the Earth's resources. A willingness to interdepend in a political climate of greater openness and stability would produce more agricultural produce at lower cost.

The main obstacle to the achievement of this scenario is political. Agricultural policies around the world reflect the attachment to the goal of self-sufficiency which in many cases has only been achieved by a high degree of protectionism. The willingness to reduce protectionism worldwide has allowed some progress to be made through the Uruguay Round of GATT talks, but not yet to the point of open markets fostering comparative advantage. Commercial interests are very strong and work against economic liberalism where domestic markets appear to be threatened. However, a growing political force which may work as a counterweight to this innate political conservatism is the concern for the environment.

The environmental arguments in favour of the growth of non-food uses are strong. Agricultural raw materials are renewable and biodegradable. The products which can be made from them offer environmentally friendly properties such as biodegradability and biocompatibility. Where they are used to replace non-degradable products that would otherwise pollute the environment, they offer a clear environmental benefit. Care has to be taken, however, in the means by which the agricultural raw materials and their derived products are produced. Intensive agricultural methods can produce their own environmental damage. These risks are now better understood and extensive methods of production can be employed in many cases. The risks do not outweigh the benefits of using renewable, degradable raw materials, particularly not when technological progress affords solutions to environmental hazards. There are many examples of the latter, but the engineering of plants which are resistant to disease and thus no longer require a high level of agrochemical protection is a good one. Such a development calls into question the extreme environmentalists' view which would prefer to see regression in the use of technology.

The pace of change is notoriously difficult to predict and particularly when so many different factors are involved. A key factor affecting the speed of growth in non-food uses is the resolve to reduce agricultural support following the GATT talks. If this resolve holds then agricultural prices could be reduced by one third, which would significantly improve the commercial viability of a number of non-food uses. Such political decisions are notoriously vulnerable, however. A change in the political complexion of the administration can produce an about-turn in policy, particularly when domestic pressure increases in adverse economic conditions.

It would be better for the policy framework for the non-food uses of agricultural raw materials to be decoupled from food policies where possible, as this would remove an additional complication in the regulation of a market unrelated to the features of the food market for agricultural produce. Such a move would meet with the approval of the industrial users of agricultural raw materials for non-food use who find the agricultural policies they have to contend with unnecessarily complicated and sometimes a positive deterrent to their investment plans. An entrepreneur is unlikely to embark on a project where the policy base upon which it depends is considered vulnerable. If non-food uses of agricultural raw materials could be regulated more simply in the policy sphere to which they relate (for example, energy policy being allowed to govern energy uses), then the choice as to whether or not to use agricultural raw materials would be reduced to a straight commercial decision as to whether or not they are cheaper and more advantageous to use. Their environmental and technical attributes will still play a role in the decision but fundamentally the entrepreneur will decide on raw material cost grounds.

The development of the oil price will play an important role in determining when the break-even point for the use of agricultural raw materials which have to compete with hydrocarbons is reached. In 1987, a group of independent consultants appointed by the European Commission concluded that either agricultural prices would have to be reduced by 40% or the price of oil would have to rise to US$40 per barrel for agricultural alcohol to compete in the energy sector. Since that estimate was made agricultural prices have fallen, although not by as much as 40%, but the price of oil despite a short period of high prices during the Gulf crisis has remained around US$18 per barrel. However, since the value of the dollar has declined in value against other currencies during this time, the price of oil has fallen in real terms in many countries; the price of oil in Japanese yen and in Deutschmarks was 40% cheaper in 1991 than it was in 1974. For the chemicals sector, there is already evidence of growth in the use of agricultural raw materials to replace oil on account of the properties, such as biodegradability, which the former afford.

The European Commission has estimated that with a continuation of present trends in economic growth and oil conservation, the price of oil would reach US$30 per barrel by 2010. The International Energy Agency, which is part of the OECD, has an energy price scenario where the price of a barrel of oil rises to $35.5 by the year 2005. It is difficult to predict the effect of the collapse of the former Soviet empire upon oil prices. If the East European countries could find a way of paying for oil, petroleum consumption could rise markedly. Western oil companies are certainly eager to bring their technology, skills and capital to bear on oil production in the former Soviet Union itself. Another important political factor is the proposed introduction of a carbon tax in Europe, which would eventually add US$10 to the cost of a barrel of oil. Therefore, one may conclude that oil prices will see some increase by the turn of the century which, combined with the fall in agricultural prices, begins to make many more non-food uses of agricultural raw materials commercially attractive to use.

This author maintains that, once uncoupled from unnecessarily onerous agricultural legislation, agricultural raw materials can be commercially viable for non-food use in the medium term. To facilitate that market opening there is a case for encouraging these opportunities by such means as tax incentives, particularly in view of the environmental benefits that agricultural raw materials afford.

INDEX

Acid rain 8, 106
Agricultural development
    developing countries 5
    stages of 4
Agrobacterium tumefaciens 63–64
Agrochemicals 21–22
Alcohol 29, 34, 37, 92, 98, 105
Allelopathic treatment 66
Animal power 4
    food for 4
    units of 7
Antibiotics 79–80, 94–96, 108
Antibodies 62, 69
Aspergillus niger 77

Bacillus thuringiensis 65–66
Biodegradability 81
    plastic 84, 94, 108
Biodiesel 101, 113
Biodiversity 22–23
Bioethanol 84, 98, 100, 127–128
    by-products 98, 111
Biogas 20, 55, 74
Biomass 101, 105, 143–145
Bioprocess engineering 62
Bioreactor 70–71
    doubling times 71
    nutrition 72, 75
Biorefinery 122
Biotechnology 22, 44
    definitions of 59
Brazil 98, 104, 113
Breeders rights 136–137
Callus growing 62
Canary grass 46
CAP 24

Carbohydrates 27–37, 12
Carbon tax 17–18, 123
Cell fusion 61
Cellulose 28, 35–36, 48
Cereal farmers 118–120
Chemical industry 27, 37–38, 108–109
    speciality chemicals 84, 94, 108
Citric acid 76, 95, 97
Commercial viability 84, 92, 113, 145
Comparative advantage 24, 146
Cosmetics 41–42
Cotton 45
Crambe 42
Crop yields 11–15, 25
Crude oil 85, 87, 92, 108
Cuphea 42
Customs classification 141

Deoxyribonucleic acid (DNA) 60
Detergents 40
Developing countries 8, 12, 18, 58
Disease resistance 61, 64, 68

ECLAIR 121–122
Electroporation 63
Elephant glass 46, 103
Energy
    balance 75
    conservation 90
    consumption of 5, 15, 16
    demand for 15–16
    diversification 6
    hydro 18
    price of 16
    renewable 9, 15–18

Index

solar 18
wind 18
Environment
   control of 10
   damage to 20, 147
Environmental Protection Agency 135
Enzymes 44, 77–78, 94, 108, 131
   immobilization of 78
ETBE 100, 107
Euphorbia 42
European Commission 36, 42–43, 90, 99, 101, 119
European Community 90–91, 101, 106, 110, 116, 131

Fats 27, 37–38
Fatty acids 38
Fermentation 69, 71–74
Fibres 27, 44–48
Flax 45–46
Food
   demand for 25
   surpluses 11
Food and Drug Administration 135
Forest resources 19
Fossil fuels 101, 145
   dependence on 7
   introduction of 5, 7
   reserves of 8, 16
France 100, 113, 132, 139
Fructose 28
Fuel oxygenates 99, 103

Gasohol 128
GATT 23–24, 92, 93, 113, 116–118, 140–142, 147
Genetic Engineering 59, 60, 134, 145
Germany 131, 138
Global warming 14, 89, 105
Glucose 28, 110
Government research funding 131
Grain surplus 128
Greenhouse effect 8, 14, 17, 23, 105
Gulf Crisis 84, 88, 106, 113

Hemp 46
Herbicide resistance 64
Hunger 1, 15, 25

Hunter–gatherers
   cultivation by 3
   energy use by 3
   keeping animals 3
   social structure 2
   use of land 2
   use of time 2

Insect resistance 65
Interdependence 24
Irrigation 14
Italy 108, 114

Japan 116, 122–123, 131, 133
Jerusalem artichoke 98

Land
   availability of 11–12
   use of 123
Lignin 35–37, 122
LINK programme 36
Lubricants 40, 42

Miscanthus 46, 103
Molasses 28–29, 75, 96
MTBE 99

Nitrates in drinking water 66
Nitrogen 77
   fertilizer 22
   fixing 22, 66–67, 145
   oxide 17
Office of Technology Assessment 59, 116, 129
Oil price 127–128, 148
OPEC 85, 88
Organization of Economic Development and Co-operation 60
Ozone layer 9

Paints 40
Paper and board 34, 46
Part-time farmers 124
Patent law 130–137

Pesticide resistance 64
Pharmaceuticals 41, 54, 112
Photosynthesis 6, 9–10
Policies
    agricultural 116, 118–121
    finance 129
    social 123
Polyhydroxybutyrate (PHB) 80–81
Population
    food requirements 10
    limits 10–11, 25
Potatoes 33
Prices, world 101, 146

Quotas 124

Rape 38, 43
Recombinant DNA 60, 67, 134, 136
Rice, 33, 125
Rural areas 107

Safety measures 79
Salinity 14
Secondary substances 52
    essential oils 53
    homeopathy 53–54
    wild flowers 53
Security of supply 106
Seed
    cost 68
    improvement 62
Self-sufficiency 24–25
Set-aside 23, 111, 118, 121, 127, 142
Soil erosion 12, 19–20
Sorghum 37, 103
Starch
    chemical structure 30
    extraction process 31–33
    grading 32
    physical properties 30–31
    regime for 109
Straw 28, 35–36
Subsidies 140–146
Sugar
    beet 98
    cane 28
    chemical structure 28
    esters of 29
    physical properties 28–29
    regime for 109
Surplus land 1–3

Taxation 89, 101, 104, 113, 128–130, 149
Technology
    intermediate 58
    old and new 59
Thermodynamics, laws of 6
Trade relations 107, 109
Treaty of Rome 118

UPOV 138
USA 98, 103, 110, 113–114, 116
USDA 135
USSR 127

Venture capital 130
Viral resistance 64

Waste products 54–55
Water
    availability of 13
    quality of 14
    rainfall 15
    wastage of 13
Wheat 32, 124
    flour 32
    gluten 32, 98
Whole crop harvesting 122
Wood 47–52
    coppicing 49–52
    fuel 19–20
    grants for 50
    harvesting 49–52
    pruning 52
    residues 22
    short rotation 48, 101, 105
    yields 48
World hunger 1, 15, 25
World prices 101, 146